No. 615
$12.95

DESIGNING & MAINTAINING THE CATV & SMALL TV STUDIO

By Kenneth Knecht

621.3886
K73d

165

TAB BOOKS
Your Best Buy in Electronics Books.

80 354 3

FIRST EDITION

FIRST PRINTING—SEPTEMBER 1972

Copyright © 1972 by TAB BOOKS

Printed in the United States
of America

Reproduction or publication of the content in any manner, without express permission of the publisher, is prohibited. No liability is assumed with respect to the use of the information herein.

Hardbound Edition: International Standard Book No. 0-8306-1615-2

Paperbound Edition: International Standard Book No. 0-8306-2615-8

Library of Congress Card Number: 72-87450

Preface

The subject of television studio system design and maintenance is very broad, due to the various demands that must be met. From the small ITV (instructional TV) studio to a commercial broadcast studio, the production capabilities vary widely. In each case, I tried to go into as much detail as possible, so little technical background will be required to understand the material presented. The material is presented in a logical sequence so that the uninitiated can start at the beginning of the book and develop his knowledge from the ground up.

I would like to thank William Stocklin, editor of Electronics World, for allowing me to use material from articles written for his publication. I would also like to thank Ronald Merrill, editor of Broadcast Engineering, for permission to reprint most of my article,"That New System—What Am I Bid?" in Chapter 14, and Merwin Dembling, editor of Photo Methods For Industry, for permission to reprint most of an article on color cameras, plus bits and pieces of other articles written for that publication.

Thanks also to my associate, Phil Livingston, for checking the manuscript for errors, and to the many manufacturers who kindly supplied me with photographs of their equipment.

Kenneth B. Knecht

Contents

1 STUDIO PULSE SYSTEM — 7
Importance of sync pulses—Various television systems—Pulse fundamentals—Sync generators—Timing pulses—Pulse distribution amplifiers—Encoders and decoders — Changeover switches — External input—Patching—Accessory equipment

2 SWITCHING SYSTEMS & SPECIAL EFFECTS — 21
Video fundamentals—Switching methods—Production switchers — Dissolves — Supers—Monitors—Remote control of switchers—Composite and non-composite video—Synchronous video—Distribution switchers—Program switchers—Bridging—Terminating—Audio follow—Special effects—Wipes—Keys—Delays

3 CAMERAS & LIGHTING EQUIPMENT — 35
Cameras—Camera theory—Lenses—Viewfinders—Camera control units—Camera accessories — Remote pan-tilt-zoom-focus — Matching cameras—Frame line generator—Camera set-up — Tripods — Pedestals — Pan-tilt-head—Test patterns—Lighting the set—Lighting instruments—Patching—Dimmers—Microphone input boxes—Studio monitors

4 THE FILM CHAIN — 57
Projection material—Slide projectors—Remote control—Animation—Film projectors—Opaque material—Splicing film—Film camera—Film island remote control—Multiplexer—Timing leader—Automation

5 VIDEO RECORDERS 73
Video tape recorders—Editing—Quad VTR—Quad accessories—Helical VTRs—Helical accessories—Video cartridge systems—Video disc recorders

6 COLOR EQUIPMENT 93
Color cameras—Encoders—Color VTRs—Phasing — Vectorscope — Differential phase and gain—Lighting—Sync generator—Color test signal—Black burst—Color picture monitor—Color adjustment standards—Color encoding systems—Color proc amps

7 RF & VIDEO MONITORS 105
RF monitors—Video monitors—Uses of monitors—Under and over scanning—Linearity—Resolution—Cross pulse monitor—Television projectors—Image enhancers—Waveform monitors—Video level meters

8 VIDEO TERMINAL EQUIPMENT 121
Video distribution amplifiers—Video test generators — Multiburst — Equalization — Stairstep—Sine-squared pulse and window—Sequential switcher — VITs — Clamp amplifiers—Processing amplifiers—Video patch panels—Sync patch panels

9 DISTRIBUTION SYSTEMS 133
Video distribution—Cable—Coax—Balanced video—RF distribution—Modulators—Passive devices—RF amplifiers—RF cable—RF distribution test equipment

10 AUDIO MIXING CONSOLES 143
Audio terminology — Audio consoles — Feedback — Cuing — Audio impedances — Bridging—A different type of console—Monitor amplifiers—Speakers—Equipment utility monitor systems

11 AUDIO PROGRAM SOURCES 155
Reel-to-reel recorders—Dolby circuitry—Cartridge tape recorders—Cassette recorders—Turntables—Tone arms—Phono cartridges — Turntable preamplifiers — Microphone directionality—Dynamic microphones—Ribbon microphones—Condenser microphones—Microphone impedance — Stands — Hand microphones — Booms — Wireless microphone — Shotgun microphone

12 PERIPHERAL AUDIO & OTHER EQUIPMENT 169
Audio distribution amplifier—Audio patch panels—Terminal blocks—AGC amplifiers—Limiters — Test tone generator — Reverberation—Echo—VU meters—Pads and splitters—Intercom system—Tally lights—Racks and enclosures—Equipment remote control

13 THE MAINTENANCE SHOP 184
Test meters—Oscilloscope—Signal generators—Component testers—Spare parts—Tools—Maintenance procedures

14 PUTTING A SYSTEM TOGETHER 198
Specifications — Boilerplate — Design philosophy — Construction practices — Equipment list—Drawing flow diagrams—Cables in racks—Cable run sheets—AC wiring—Grounding—Cables—Cable routing in a room or building—Pulling cable

15 THREE SYSTEMS DESIGNS 209
The single camera-VTR-monitor system—Equipment required—Physical and electronics layout—Choosing specific equipment

INDEX 250

Chapter 1
Studio Pulse System

Television studio systems can be divided into four basic types: the commercial or public television studio, the production company studio, the instructional TV (ITV) studio, and the CATV system studio. Commercial and public television studios are essentially the same. Their task is to originate programs—live, video taped or on film. The major equipment usually found includes quadrature (quad) video tape recorders (VTRs), film chains, studio cameras, production and program switcher, and audio mixing console. The programs are fed to a television transmitter to be put on the air for potential viewing by any interested TV set owner. Fig. 1-1 shows a large master control room in a commercial network center.

Fig. 1-1. NBC master control room in the RCA Building in New York City. (Courtesy NBC)

Fig. 1-2. Typical ITV studio showing cameras on pedestals, control console and lighting.

The production studio is set up to record commercials or programs which are then distributed as desired by the people for whom the material was recorded. The production studio uses the same equipment as the commercial studio, with the exception of the program switcher and transmitter. The video tape equipment is usually more elaborate, sometimes better quality machines are used for "mastering" (recording "master" tapes from which "dubs," or duplicates or made) and more machines are available for tape duplication. Sometimes high-speed duplication equipment is used instead of "dubbing" on standard machines in real time. Usually, a video disc recorder is also present, available for slow and stop motion effects when required. Some of the large public TV and commerical stations also do outside production work, too.

ITV studios range from a simple system using limited equipment to one that rivals the commercial station. An example of a typical studio is seen in Fig. 1-2. A distribution switcher is usually desirable so that more than one program can be originated at the same time. The ITV system uses one or more modulators and a cable distribution system to reach its audience. Programs are routed to selected classrooms to be used in the teaching process.

The CATV system studio can range from a simple oscillating camera showing time and weather data, plus other information successively, to a more complex studio such as that used in a large ITV system. The studio is used to supply programming to one or more unused channels on the cable for viewing by system subscribers.

All the equipment discussed so far is described in much more detail later. We also cover other necessary equipment not yet mentioned. The quality of the equipment discussed ranges from the inexpensive models which can be used by the small ITV or CATV operator through the top-of-the-line equipment used in the largest studios. However, the accent is on the smaller systems. Sample system designs—medium and large—are illustrated in the last chapter. Simple RF distribution systems, maintenance and test equipment, plus other peripheral systems are also covered.

In any TV studio several types of pulses are necessary for the generation of a TV signal. It is essential that these pulses occur at precisely the correct time with relation to each other and the video. For example, if the scene viewed by the studio camera is to be an exact reproduction on the screen of the final TV set at the end of the system, the various pulses all must be positioned correctly, otherwise the picture will be distorted. Of course, the video information must be as precise an electronic reproduction of the original scene as possible. When we say the signal at the final TV set must be an exact reproduction, we mean within the electronic limits of the equipment used. This is discussed in more detail when we describe individual equipment in the system.

PULSE GENERATORS

A pulse is an abrupt change in voltage from one potential to another and back again. The width of the pulse is the length of time the voltage is deflected from the original point. For example, let us assume our voltage is at 0 volts, then is electronically switched to -4 volts for a second, then returned to 0 volts. This results in a negative pulse 4 volts in amplitude with a width, or duration, of one second. As you can see, pulse amplitude is measured in volts or fractions thereof and pulse width is measured in seconds or fractions thereof.

In the case of sync generator pulses, the amplitude is typically minus 4 volts out of the sync generator, with widths varying in accordance with the pulse train that is being considered. A pulse train is a series of pulses, sometimes alike, sometimes different, but always a repeating series (in the case of sync generator pulse trains).

The speed with which the pulse changes in amplitude determines the shape of the pulse. In the case of sync pulses, this change is very abrupt, resulting in very steep-sided pulses, so steep that they appear vertical as represented in an illustration or seen on an oscilloscope. Of course, if the pulse is illustrated in extreme horizontal or time magnification, some slope will be apparent. For all practical purposes, however, the pulses will appear to have vertical sides. Such a representation is called a waveform. Any representation of an electronic signal, whether an illustration or an oscilloscope pattern, is called a waveform.

Every television system, from the simplest to the most elaborate, requires a sync generator of some sort. The broadcast (EIA) sync generator is used in all but the smallest systems. When we use the term EIA (Electronics Industry Association), it should be taken to mean that the generator meets the EIA RS-170 specifications for broadcast sync. The following pulse trains are supplied by this type of generator— composite sync, horizontal drive, vertical drive and mixed blanking. As will be seen in a later chapter, a color sync generator puts out some additional signals which are required for color equipment. A photograph of a color sync generator appears in Fig. 1-3.

Composite Sync

Sync consists of horizontal pulses, equalizing pulses, and vertical pulses. It is used to maintain the proper scan to recreate the original scene on a monitor or TV set. It is also used to keep all the video sources operating on the same time base. This allows the switcher operator to fade from one source to another or "take" between them with no rolls or other disturbance in the resulting picture. Almost all the video equipment we discuss requires sync or one or both of the drive signals in step with the sync.

The horizontal pulses separate each line of video information from adjacent lines. These pulses are used to initiate the return of the beam in the picture tube to the left side of the screen to start another line. The horizontal sync frequency is 15,750 pulses per second.

The vertical pulses in the composite sync signal time the vertical scanning interval. They cause the beam in the picture tube to return to the top of the picture and resume the horizontal scanning. The vertical sync is cut into pulses one half line in width by what are called serrations. Instead of leaving the vertical sync one long pulse, it is serrated so that horizontal scanning can continue while the beam is returning to the top of the picture. If the horizontal timing were lost during vertical sync, the top of the picture would be unstable until the horizontal oscillator in the TV set locked in with the horizontal sync again. The amount of disturbance is determined by the AFC (automatic frequency control) and the accuracy of the setting of the stability control in the TV set horizontal circuitry.

Equalizing pulses precede and follow the vertical serrations in the vertical interval. The purpose of these pulses will be clear shortly. The lines between each vertical interval are called a field. Two fields make one TV frame. The frame, or information required to make a full TV picture, is divided into two fields to reduce the flicker which would otherwise be caused by changing all the picture information every 30th of a second. By changing half of it every 60th of a second the persistence of the image in the eye eliminates the flicker. The halves of the picture are interlaced together; alternate lines are supplied by successive fields. If one field starts with the first line at the left edge of the picture, the next field must start its first line in the center of the picture to provide the proper interlace. The equalizing pulses provide this interlace function. They occur at twice the horizontal sync rate, as do the

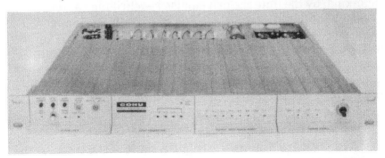

Fig. 1-3. Broadcast EIA color sync generator. (Courtesy Cohu)

vertical serrations. Because of the half line preceding one field and the half line following the next field, alternate equalizing pulses and serrations provide the horizontal sync during each vertical interval. The equalizing pulses also provide an equal charge in the vertical integrating capacitor in the TV set regardless of which field is being processed. The vertical intervals for each field are shown in Fig. 1-4. Notice where the half lines fall in each field.

The half line occurs because the fields are 262.5 lines in duration. This is due to the fact that a frame consists of 525 lines (525 divided by 2 equals 262.5). The number 525 comes from the fact that when the standard was originated it was easiest to make up the standard from small integers, e.g. 7 x 5 x 5 x 3 equals 525. This made it easy to design a sync generator using electronic multiplier circuits available at that time.

Horizontal and vertical blanking pulses are wider than the respective sync pulses and are used to blank the video a short time before and after these pulses. See Figs. 1-4 and 1-5 for comparative durations. This blanking is required to make sure there is no video in the picture when the electron beam in the picture tube is returning to the left side or the top of the picture. If this were not so, the picture would be degraded by extraneous video and zig zag retrace signals during the retrace interval. Both horizontal and vertical blanking are supplied in a single pulse train, called mixed blanking.

The horizontal and vertical drive pulses are about half the width of the respective blanking pulses; their start coincides with the leading edges of the blanking pulses. This is also illustrated in Figs. 1-4 and 1-5. This difference in width is to allow for delays caused by long camera cables or other interconnecting cables. If the pulses were the same width as the blanking, by the time the drive pulse reached the camera over a long cable, initiated a line, and the resulting line of video returned from the camera, the video might start after the blanking pulse ended. This would put a black bar, with no video, at the left side of the picture. In the case of the vertical pulses, the black bar would be at the top of the picture. Granted, it would take a very long cable to cause a significant delay, but think how far away the cameras might be on a sports remote, such as a golf match.

The drive pulses are used to time the cameras so the signal originated is in step with the sync pulse train. Most

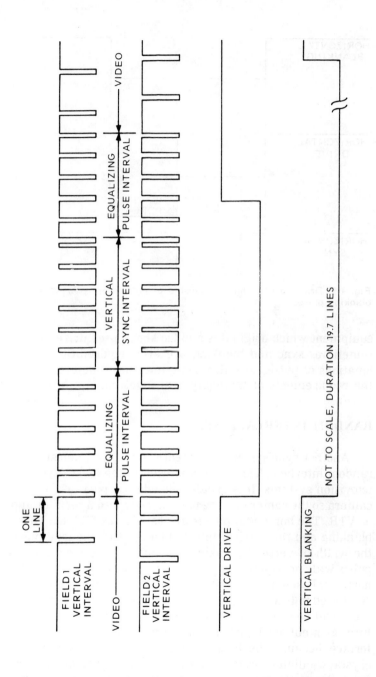

Fig. 1-4. Pulse train showing vertical interval pulses for each field, plus drive and blanking pulses.

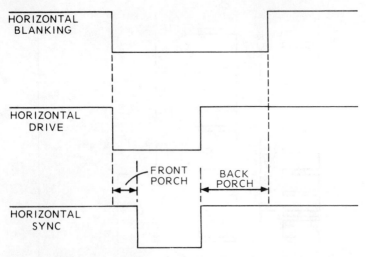

Fig. 1-5. Drawing showing relationship of horizontal sync, drive and blanking pulses.

equipment which originates a video signal uses drive pulses; sometimes sync and blanking are also required. Other combinations of pulses are also sometimes used, determined by the requirements of the equipment involved.

RANDOM INTERLACE SYNC

Another sync system in current use is called industrial or random interlace. This system is used in the least expensive television systems. It is usually included as part of a vidicon camera so the camera can be used alone to feed a monitor and-or VTR. The horizontal pulses are as wide as EIA horizontal blanking and the vertical interval consists of one long pulse, the width of vertical blanking. Because of the long vertical pulse with no equalizing pulses or vertical serrations, the horizontal oscillator circuitry in the TV set or picture monitor is allowed to free-run during the vertical interval. This means there will be some horizontal instability at the top of the picture, as mentioned earlier. This system is called random interlace because the horizontal pulses are derived from a crystal oscillator circuit and the vertical pulses are derived from the 60-Hz power line frequency. Since the two pulse trains are derived from different standards, they tend to drift back and forth with relation to each other. Since interlace depends on a stable relationship between the horizontal and vertical pulses, the lines are sometimes interlaced, sometimes overlapped in the resulting picture.

Oscillator circuits can be designed to provide a signal with any pulse repetition rate desired. Usually, a quartz crystal is used to ensure the accuracy of the desired frequency. If the crystal temperature is kept constant, it will reliably hold the oscillator to its designed frequency. This signal is then used to provide the pulse generating circuitry with timing information. The pulse generating circuitry can also be timed with an external signal, such as the 60-Hz power line frequency to provide the 60-Hz vertical sync pulses.

Thus, in the random interlace case the horizontal scanning frequency is provided by a 15,750-Hz crystal oscillator which supplies the timing information, while the 60-Hz power line times the vertical pulse generating circuitry. To repeat: The horizontal pulses determine the number of lines scanned during a given period of time and the vertical pulses determine the number of fields produced in a given time.

PULSE DISTRIBUTION

The pulses are 4 volts in amplitude when they leave the EIA sync generator and they can then be fed directly to the various pieces of equipment. However, if a piece of equipment must be disconnected for repair, any equipment "downstream" from that point is deprived of the pulses until the two cables are reconnected. Also, if the pulses are fed to many pieces of equipment the waveform tends to become degraded by the different loads. This, of course, is based on the assumption that the pulse inputs are bridging, not terminating. This is discussed further in the next chapter.

To avoid the above problem, unity gain amplifiers with multiple outputs, usually four or six, are available. These amplifiers are designed to handle all the different pulse trains without degrading the waveshapes. Sometimes these amplifiers just amplify the signal enough so it can be divided, but usually the pulse distribution amplifier (PDA) is designed so that it regenerates the pulse. This is useful if the pulse has been carried over a long line and its shape has been distorted. A gain control is usually supplied to adjust the PDA outputs to exactly 4 volts. Fig. 1-6 shows a modular pulse distribution amplifier removed from its frame.

Each PDA output can feed a different group of equipment. If any one of the output lines is disconnected, only the equipment fed by that line will be affected. In a system perhaps the critical equipment can be provided with pulses from separate PDA outputs, with the monitoring equipment grouped together on another PDA output. The non-critical monitoring equipment only will be disturbed then if the connection to one of a group of equipment is interrupted; the critical equipment will remain undisturbed on the other PDA outputs. Sometimes, however, there will be too many pieces of critical equipment to supply with one PDA if this scheme is used. In this case you might consider using an extra PDA to provide the additional outputs. A pulse fanout module, described later, could also be used.

An alternative to using PDAs for each pulse train is a system of encoders and decoders designed to accept all outputs from the sync generator, encode them into one signal which is fed to the decoder, where it is changed back to the pulse trains originally fed to the encoder. Thus, all the pulse trains can be carried over one cable, delayed if necessary, and then decoded at the location where the pulses are required. This can eliminate many PDAs, since a single encoder can feed many decoders. The delay is used where pulses are fed to equipment at the end of short cables, so those fed to equipment over a long cable will arrive at the switcher in phase with all the others. This system is obviously more important in large systems. It is best to calculate the pulse system price using both systems. If you had planned a lot of pulse patching and PDAs, the costs will be about the same. A big advantage of the encoder-decoder system is that all the equipment fed by that decoder can be switched to another sync source (encoder) by merely patching one cable. This would be much more complicated if only PDAs were used, requiring each PDA to be transferred to another source separately. Of course, you can use PDAs on decoder outputs if desired. However, a PDA should not be used to split up an encoded signal, since it is not designed to process a signal of that waveshape and amplitude.

An interesting and useful addition to a pulse system might be a sync changeover switch. To use it you need two sync generators, one for the pulse system and the other as a standby unit. Both generators are fed into the changeover switch, which has two sets of outputs. One set, called the primary output, is used for house sync; the other, called auxiliary, feeds a patch panel so that it can be used for anything you desire. More about sync patching later. The sync changeover switch can be manually operated with optional remote control. The changeover can be operated automatically, too; if any pulse train deteriorates in amplitude it automatically changes all its inputs over to the other generator and turns on a warning light. This lets you know you are operating on the backup generator. Of course, you have then lost the auxiliary outputs, which have been switched to the defective generator.

Sync generators themselves are of various types. Most of the newer generators produce digitally derived pulses; that is, digital counters are used to form the pulses. A generator of this type needs very little attention, and pulse widths do not have to be continually checked and adjusted.

Most sync generators allow you to lock them to various timing standards. One such standard is line lock, in which the generator derives its time base from the 60-Hz power line frequency. Unlike the random interlace generator, the horizontal pulses are also derived from the 60-Hz signal. This is very adequate for monochrome recording, editing and playback. For color, a crystal lock position is available on some generators. This, as the name implies, uses a very accurate crystal oscillator usually employing a crystal oven to avoid thermal effects. A third standard is called external or remote, which allows the generator to be locked to any ex-

Fig. 1-6. Pulse distribution amplifier module. Several plug into a rack frame. A separate power supply module also plugs into each frame. (Courtesy Dynair)

ternal standard. For example, you could feed NBC programming into this input, assuming NBC to be your network affiliation if you are a commercial station. All your equipment is now locked to NBC, so you can super your ID (station identification) over them or dissolve from network to a local commercial.

Sync patching, if used, amounts to bringing the inputs and outputs of all the PDAs, the outputs of the sync generator or changeover switch, and the lines feeding the equipment to patches, so that the pulses can be rerouted if necessary. If a spare PDA or two is also brought to the jackfield, it can be quickly substituted for an ailing PDA by patching around the defective unit. We discuss patching in much more detail later. The equipment used is identical.

Other accessories for handling pulses are also available. For example, a pulse fanout module can be used to split a pulse line into several outputs. It differs from a PDA in that it only splits the signal; it does not reshape the pulses. It usually has more outputs, too. It is meant to be used following a source of well formed pulses, such as a PDA or sync generator. It should not be used at the end of a long connecting cable because long cables tend to distort the pulses. Fig. 1-7 shows a pulse fanout module removed from its rack frame.

A sync generator with built-in delay circuitry can also be used to delay pulses. For example, you could feed such a generator with a sync pulse train and the generator would then provide delayed sync, blanking and drive signals. In special cases, perhaps you might wish to feed sync, blanking and drive to a group of equipment at a distant remote location and wish to keep it synchronous with the signals at the source. You could run four cables to PDAs at this location, but possibly due to the distance you might wish to use only one cable. You could feed sync over that one cable to another sync generator's remote input, and let it then provide the required pulses. This is even more attractive in a color system, which would require six cables instead of only four. In an emergency, if the incoming sync is lost, you would still have pulses if the generator were switched to line or crystal reference. Of course, its output would no longer be synchronous, but at least you would still have all the pulses at the remote point. Some generators will switch to another reference automatically if the external input signal fails.

Another special sync generator will provide a complete output pulse array even if there are pulses missing in the sync fed to the external input. A generator like this is useful when the external input signal is provided by a helical VTR using only one head. With this type of VTR, pulses are missing in or near the vertical interval. The reason for this is discussed in the chapter on VTRs.

Special PDAs that will delay any input are also available. The need for such a delay was discussed earlier. Other PDAs will eliminate any extraneous signals superimposed on the pulse train, such as 60-Hz hum induced on a long interconnecting cable by power lines running near them. This feature is provided by clamping circuits. Clamping circuits are discussed in more detail in another chapter.

The sync generator can also provide other pulses such as subcarrier, burst flag and crosshatch. The subcarrier and burst flag are discussed later. When a sync generator is used

Fig. 1-7. A pulse fanout module. It is used to obtain extra outputs from a pulse distribution amplifier. (Courtesy Grass Valley)

in the external mode and the source is switched to another nonsychronous signal, three things can happen, depending on the generator. Its output might become very disturbed until it locks in with the new input. It might just automatically add or subtract a line or two a field until it becomes synchronous with the new pulse train. This means it would retain horizontal synchronization and regain vertical synchronization in two seconds at the most, losing only vertical synchronization during this period. The third mode allows it to regain vertical lock as soon as the correct vertical interval appears, a 30th of a second at the most. Of course, as soon as it regains vertical synchronization with the correct field, it would also regain horizontal synchronization.

The best mode to use depends on the circumstances. In most cases the third mode would probably be the best. Possibly if recording or playing back color on a quad VTR, the second mode might be preferable. A little experimentation to see which method gives the least disturbance to the picture would be best in most cases.

Chapter 2
Switching Systems & Special Effects

As an aid to understanding the various switching methods used in a TV studio, we should begin with a definition of a video signal. Video is the term used for the signal that carries the picture, blanking and sometimes the sync information. Assuming the video signal, including sync, is typically 1 volt in amplitude, if this voltage is divided into 140 equal divisions, as is usually done, 40 units, or IRE divisions, represent sync, 10 units the setup or pedestal and 90 units the picture information. The signal is measured in the following manner: the bottom or tip of the sync is referenced to -40 IRE units, the top of the sync pulse is called 0 IRE, the bottom of the video signal is +10 IRE and the top (peak) video is +100 IRE. The setup is the part of the video signal between the bottom of the picture information and the top of the sync pulse. Sometimes this is called blanking amplitude.

The relationship of the video signal to the picture information is quite simple. The video at +100 IRE units is white; at +10 IRE units it is black. Any amplitude between these two points is a shade of gray. The more vertical the transition of the signal between two shades of gray, the greater the resolution or detail seen in the picture. It can be seen best in a transition between black and white. This resolution is a function of the characteristics of the camera image tube or other signal source and the bandwidth or frequency response of the amplifiers in the system.

SWITCHING METHODS

Now to the switcher. We will begin with the production switcher. The output is usually fed either to a VTR or is used for live origination.

In elementary form, the switcher is a row of pushbuttons, each connected to a different video input, all feeding a common output. Only one button at a time should be pressed. You can thereby "punch up" any input to the output as desired. The inputs are usually "Black" (sync and setup, no picture information), the cameras, the film chain cameras, the VTRs, the video test generator, the special effects, and any other video inputs desired. The devices used to supply these signals are discussed later in much more detail. Such a simple switcher is shown in Fig. 2-1. The pushbuttons are mechanical switches. They switch the video directly through the mechanical contacts in the switch or the contacts can activate an electronic switching circuit. Collectively, the electronic switching circuits are called the switching matrix.

One of two types of electronic switching can be employed—fast lap or vertical interval. The fast lap is just what the name implies—a fast fade from one input to the other. The lap occurs so rapidly it looks like a direct switch. The vertical interval switch is a very fast switch that is timed to occur between the fields of video, during the vertical interval when no video (only sync pulses) is being passed through the switcher. The vertical interval switch is presently the most popular for production switching. Almost all manufacturers now use this method.

The reasons for the various types of switching are as follows. The mechanical switch is the cheapest by far, but due to the physical construction of the switch either there will be a short overlapping of the signals while one pushbutton is pressed and the previously pressed button is mechanically released, or there might be an absence of signal if the previously operated button releases before the button being pressed has made contact. In any event, extraneous signals, called transients, are generated when the contacts make and break. These transients are added to the video signal, and can

Fig. 2-1. This is a simple mechanical 6-input, single-output video switcher. (Courtesy Dynair)

raise havoc with some following equipment; for example, VTRs and picture monitors.

The fast lap was popular because it used simpler and, therefore, less expensive electronic switching circuits. Because of the fast lap switching interval, overlap of the full voltage of the two video signals or an absence of signal during the switching interval were avoided. Switching could occur anywhere in the video field without causing transients.

The vertical interval electronic switch, commonly used now, is the fastest and theoretically causes the least disturbance to the picture, since switching occurs during the vertical interval. Any transients that might possibly be generated fall in the vertical interval, where they can have no effect on the video signal.

The buttons used in the switcher are of two basic types. One kind stays down when pressed and is mechanically interlocked with the others. Thus, it releases when another is operated. The other type is of the momentary contact variety that does not remain down when operated. Both types usually are illuminated so that the operator can tell at a glance which button was last pressed. There is little advantage in the use of one type over the other, except the momentary type lacks the mechanical interlocking characteristic which eliminates a possible source of trouble in the future. However, this is usually unlikely. The momentary contact switch also has a better feel, which some operators prefer. However, if the lamp used to illuminate it fails, it is hard to tell which button was pushed last.

TYPICAL SWITCHERS

Each row of pushbuttons is called a bank. Production switchers can have from one to eight (sometimes more) banks of buttons. For a starter, let's take a look at a small 3-bank switcher. All three banks are located directly under one another and have the same inputs. The top bank is usually called the preview bank and allows an input to be viewed before switching it on the line. The preview bank feeds the preview picture monitor and can also be used for other purposes in a larger switcher. More about this when we discuss special effects and camera external viewfinder inputs. We now have two banks left. These two are usually connected to

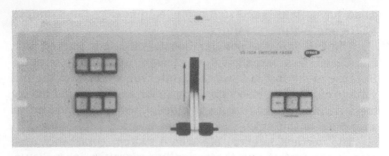

Fig. 2-2. A 5-input 2-bank production switcher with the ability to mix three of the inputs. (Courtesy Dynair)

the output line through a mix amplifier. The mix amplifier allows the operator to fade or dissolve between the two banks. A lever is mounted in the switcher panel to the right of the two banks. When the lever, called the fader, is moved from the upper bank to the lower, the upper bank video fades out as the lower bank video comes up in amplitude. Due to the circuitry in the amplifier, the signals are added together through this cycle to keep the output at its normal amplitude. The effect is a smooth dissolve from one input to the other. By leaving the fader lever in, let's say, the bottom position, the lower bank can be used as a "take" bank. This allows the operator to cut directly from one video input to another, utilizing the switching circuitry discussed earlier. Many smaller systems use a switcher like this.

Another type of switcher uses only two banks with a fader between them—no preview bank **per se**. In some versions of this type, the bank that is not selected is automatically applied to the preview monitor. That is, if the fader is pushed towards the top bank, the bottom bank will automatically appear on the preview bank. For some types of programming, this works very well; it is easy to operate and costs less than a 3-bank switcher of equal quality.

Fig. 2-2 shows a small 2-bank switcher. The fader lever is actually two levers, which can be locked together to act as one. You will notice that when the fader levers are in the mid position between the two banks you have an equal amount of each input on the output line. Let's suppose one input consists of the word "tree" in white letters on a black background and the other the image of a tree; the resulting signal on the line would be a tree with the word "tree" superimposed (supered) over it. However, if you look at the signal on a waveform monitor, which gives a graphic representation of the signal, you will see neither signal is at its full amplitude. If you split the fader arms, and cheat them away from each other a little, you can bring both signals up to full amplitude. In practice, you would probably only cheat the lever controlling the letters, leaving the other at the center position. This would make the white letters stand out from any white in the other image. You have to be careful not to raise the white letters over the proper amplitude (100 IRE units), or you will get dark streaks in the picture following the letters and a nasty buzz in the TV audio. This is caused by negative overmodulation of the picture carrier in the transmitter or modulator. A graphic representation of the supering process is seen in Fig. 2-3.

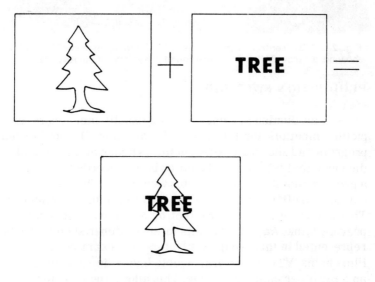

Fig. 2-3. Two images can be supered (superimposed) to produce one picture.

Usually, letter supers are done by keying, if you have a special effects unit, but they can be done by mixing as described. We discuss keying when we get to special effects. If you use a little imagination you will be able to dream up a lot of uses for this technique. A careful scrutiny of commercial television programs, especially variety shows, will reveal many possibilities.

Fig. 2-4. NBC control room in the RCA Building in New York City, showing a large switcher-picture monitor system. (Courtesy NBC)

PRODUCTION SWITCHER

The production switcher position usually includes several picture monitors for the use of the director. One is labeled program and shows the output of the switcher program bank—the video that is going on the air or being recorded. Another is a preview monitor, fed by the preview bank. The various mix and special effects amplifiers can also feed separate monitors. This allows setting up an effect or super without using the preview bank. Any other inputs to the switcher should also be represented in this group of monitors; for example, cameras, film chains, VTRs, network, remote line, video test generator, and any other inputs available. This allows the director to see the picture he will be taking next. The monitors are usually equipped with tally lights so the director is reminded which one is on the air without looking down at the switcher buttons. In a very simple system, using only two cameras and no special effects, the camera not on the air is punched up on the preview bank. This means only two monitors are required—program and preview. A large monitor system of this type is shown in Fig. 2-4. More banks can be added, perhaps two more banks for special effects and a take bank. The take bank is used for direct switches between sources without using one of the mixer banks. We go into more detail about these when we cover special effects and switcher operation.

Physically, switchers come in two versions; one type is self contained, with all connectors, electronics, and the front panel with all the pushbuttons, fader arms, etc., in one unit. This means that all the input cables have to be routed to the production position. This can make things a bit cluttered if a large number of cables are required. The other version is remote controlled, with a single, small multiconductor cable carrying the DC switching information between the control panel and the electronics section in an equipment rack. The remote controlled version is usually a bit more expensive, but if the system is large enough to employ distribution amplifiers and video patching, it would be worth thinking about.

There are two ways video can be fed to the switcher—as a part of the composite signal or in noncomposite form. The difference between them is that the composite video comes complete with sync pulses, while the noncomposite signal lacks them. Some switchers will handle either type, others require either one type or the other, and a third type will also take either type but requires a switch to be set to a position corresponding to the video type used. There is a separate switch for each input; sometimes one switch determines the type of video acceptable to all the inputs. In other versions of the third type only some of the inputs are selectable; the others are usually restricted to composite video.

Another consideration is synchronous and nonsynchronous video. Video signals are synchronous when the output of two or more video sources derive their sync or drive from a common source. You can dissolve between synchronous video sources, or take from one to the other without any distrubance in the picture. This cannot be done with two nonsynchronous sources.

Some switchers have warning indicators that alert you to the fact that the input you have selected for your next source is nonsynchronous with the house sync. Let's say you have a source selected on one of the mix banks. If you punch up a nonsynchronous source on the other bank the warning indicator lights. If you insist on dissolving, most switchers so equipped will not allow it, but will take from one mix bank to the other as the fader is moved from bank to bank. This is accomplished by using sync comparator circuits.

DISTRIBUTION SWITCHER

The distribution switcher operates in a manner similar to the production switcher; identical switching techniques are used—mechanical, vertical interval or fast lap switches **via** local or remote control. The switcher is usually used for the distribution of composite video, but most will also work with noncomposite video. This switcher does what its name implies—it allows selection of one of many inputs and routes it to one of many outputs. For example, a 10 by 10 switcher has ten input and ten output lines, with 100 buttons. Each vertical row of ten buttons is connected to the same input and each horizontal row of ten buttons connects to an output line. You can buy these switchers in many numerical configurations. Usually, you are restricted to set numbers of inputs, depending on the manufacturer, but you can get as many output buses as you desire. The remote control version can be the most flexible, allowing you to put each panel where it is to be used. For example, you might put a single bank panel at each of your VTRs, allowing you to select any desired source right at the VTR, avoiding a trip to the equipment racks, which might be in another room, to change inputs. The distribution switcher switches audio as well as video, unlike the production switcher which switches only video. Such a switcher can be seen in Fig. 2-5.

PROGRAM SWITCHER

The program switcher is like a production switcher with audio follow. It is usually used in a broadcast station that keeps programming on a single line all the time. Usually, different audio sources can be selected to follow the selection of a given video input. For example, during a station break, an identification slide might be accompanied by live audio from the announce booth or the ID audio might be recorded on an audio tape cartridge. Frequently, the program switcher is automated. The automation can be as elaborate as you desire. For example, it can be controlled by a computer or simply constructed using stepping relays and an operate button for each event.

TERMINATED & BRIDGING INPUTS

To return to switchers in general, some use terminated audio and video inputs, while others are bridging types. The terminated switcher uses a resistor to terminate the input when it is not selected; the device connected to the output terminates the selected source, of course. This means that the switcher has to be at the end of the audio or video line. Therefore, each bank of the switcher requires separate sources; usually, a distribution amplifier is then required so that a single source can be fed to other switchers or other loads. The advantage is that inexpensive mechanical switchers can be utilized, if the resulting switching transients can be tolerated. This expense has to be balanced against the possible need for distribution amplifiers.

Another version of the terminated input switcher has 75-ohm inputs to an internal amplifier. This type does not cause transients on the line when it is operated as does the purely mechanical unit.

If more than one or two switcher banks employ the same input, a bridging type switcher would be the better choice.

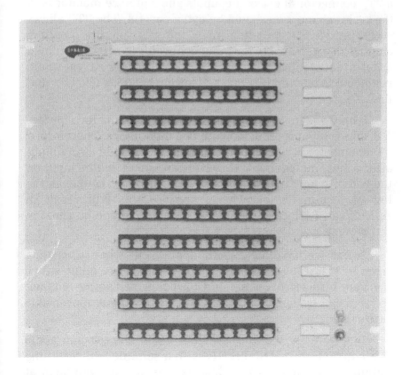

Fig. 2-5. A 12-input 10-output audio-video locally controlled distribution switcher. (Courtesy Dynair)

This allows the source to be looped through the input, because in a well designed switcher the video is not affected, permitting the same video line to be fed to many switcher banks. The bridging switcher usually uses built-in amplifiers, similar to DAs, both for the audio and the video inputs. There are passive (no amplifiers) switchers available that have bridging inputs. However, the output load has to be very close to the switcher output in the passive type. Also the load the switcher is connected to must be a bridging type, too; it cannot be terminated.

The subject of terminations is very important as it can cause no end of problems. First, let us discuss video again for a moment. A video amplifier is designed to feed a 75-ohm load. If the load is greater than 75 ohms (for instance, an unterminated bridging input), the video will be much too high in level. If the amplifier is double terminated (two 75-ohm inputs across the same line), the level will be much too low. If you use a "T" connector at one of the inputs and a picture monitor is in the series string, you should be extra careful. The "T" allows you to bridge a terminated input and most picture monitors employ a switch next to the input connector to select a bridged or terminated input. Therefore, you can see how easy it would be to double terminate a line and lower the video level on that line.

If only one input jack is provided, the input is terminated. A bridging input is provided with two input jacks. Therefore, a "T" connector is used only if you want to feed a bridging input after a terminated input. However, an unterminated input on the end of a long line should also be avoided, even though the line is terminated elsewhere, because this will result in reflections. Reflections resemble a double image or ghost in the picture.

So far we have not exactly spelled out what a bridging input is. It is a high-impedance, low-capacitance input, which has very little effect on any line it is connected across. In fact, several can be connected across one line without appreciable disturbance.

In a switcher that also switches audio, the 600-ohm audio line can be terminated directly with a suitable resistor at the output of the equipment feeding the line, or the equipment at the end of the line can supply the load. If all the inputs can be bridging types (approximately 20,000 ohms), then it is much simpler to terminate the audio at the source. This allows other inputs to be bridged across the line as desired, or you can break into the line in the patch panel with no worry about changing the audio level on the line. Unfortunately, it is sometimes difficult to find certain types of equipment with bridging inputs. You cannot simply use a bridging transformer because it would lower the level by 20 db.

SPECIAL EFFECTS

The special effects unit has various uses in television. It can be used as an instructional tool or an attention getter. However, before we talk more about special effects it might be well to define a few terms.

For example, a wipe is a vertical line that moves across the screen from one side to the other. In so doing, one image replaces the other. There are many kinds of wipes: vertical, horizontal, corner, or fancier ones such as a diamond or circle that grow from the center of the screen. Other wipes move across the screen like a windshield wiper blade. The wipe position is controlled by rotating a knob or by a lever which looks like a mixer fader arm.

A key is something like a super, but it allows many other possibilities. You can select any portion of a picture that is lighter in color than the rest of the picture and superimpose it over another picture. If you desire, you can then insert a third video source into the keyed in portion. For example, let's take a picture of the tree example used earlier. Another picture can be a small square on a darker background, which can be keyed over the tree. This square can be made any color by the keying circuit, from black through shades of gray to white. If we want to, by using the external keying input, a third input (for example, a picture of a seed cone of that tree when positioned so it will fall in the area covered by the square) can be inserted in that square. You can think of many other possibilities, I'm sure. The above sequence is shown in Fig. 2-6.

As you can see, the use of special effects requires that three additional banks be added to the production switcher, although sometimes the preview bank is used as the external keying bank; therefore, only two additional banks are required. These additional banks can be part of the switcher, along with the special effects unit, or three distribution switcher banks can be fed to the special effects unit, the output of which is fed to the switcher as another input. If you use this scheme you will run into a small problem. The video that passes through the special effects is electronically delayed with respect to the other inputs to the switcher because of the delay in the special effects amplifier. This means the output of the special effects is no longer synchronous with the other video inputs.

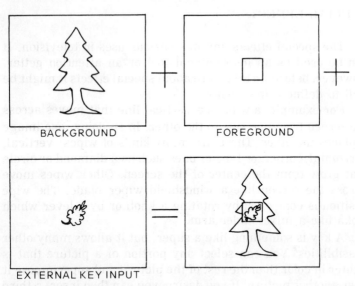

Fig. 2-6. Three images, including an external key input, are used to produce the final image shown.

There is a simple solution to this problem: all you have to do is delay all the other inputs to the production switcher by an equal amount. This delay can be achieved three ways. If the effects unit is built into the production switcher, the delay compensation is taken care of. If you added the special effects yourself, you have two choices. You can use cable, which has an inherent delay, or special lumped delays which are physically much smaller than a coil of cable. In practice you loop the inputs through the three added banks feeding the special effects, then run each input through a delay unit of either type, then to the production switcher. The effects output is run directly to an unused switcher input. If you use cable or adjustable delays, just set the fader halfway between the effects and another synchronous switcher input and trim the length of the coil of cable or the adjustable lumped delay until the edges of the horizontal sync or blanking pulses are superimposed as seen on an oscilloscope. Then adjust all the other delays for an equal amount of delay. If cables are used, the other delay cables just have to be cut the same length as the adjusted one. If you are switching monochrome the delays need only be very close, but if color is used the delays are very critical.

You can hit the required cable length pretty closely if you use the following table. The series is linear, and can be extended as far as you like. To use it you look up the delay of your special effects (listed in its specifications or obtainable from the manufacturer) and cut the cable to a length equal to the delay time required. It might be wise to cut it a bit long and trim to the exact length using the scope.

```
1 nsec—7.9"         6 nsec—47.2"
2 nsec—15.7"        7 nsec—55.1"
3 nsec—23.6"        8 nsec—63.0"
4 nsec—31.5"        9 nsec—70.9"
5 nsec—39.4"       10 nsec—78.7"
```

You can use as many effects units with a switcher as you desire. Some switchers use eight banks: a preview bank, a take bank, and three other pairs of banks. Each of the three pairs can be used as mix banks or effects banks by pressing the appropriate button next to each fader handle. Each pair can feed into the other two pairs as well as the preview and take banks. The preview bank can be used as the third bank for any of the pairs, thus allowing external keying. You can see the possibilities. For example, you could dissolve a key in over an effect and then dissolve that into another input; or you could wipe behind a key and dissolve that into another input. The possibilities are almost endless. If you get a switcher this large, it is considered a custom type, and you can include any features you desire. When we discuss color, more possibilities will come to light.

Of course, delay problems become more complicated in such a complex switcher, but they will be taken care of by the manufacturer. An example of a large production switcher is shown in Fig. 2-7 and the modules in Fig. 2-8.

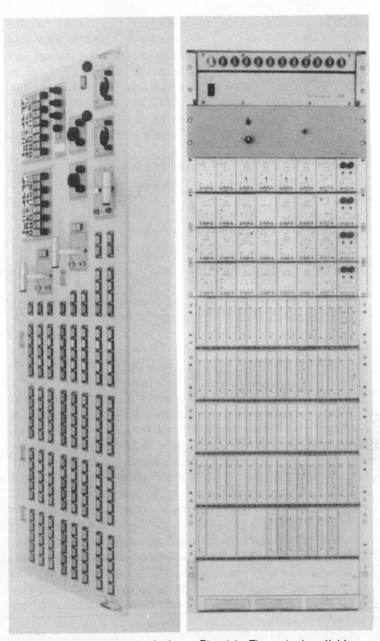

Fig. 2-7. A large 8-bank dual-effects switcher with "all the bells and whistles." (Courtesy Grass Valley)

Fig. 2-8. The actual switching modules and other electronic circuits for the switcher shown in Fig. 2-7. (Courtesy Grass Valley)

Chapter 3
Cameras & Lighting Equipment

1739244

The shooting stage or studio is an area containing the cameras, sets and lighting necessary for a video production. In this chapter we consider only monochrome operations. Color is taken up separately.

The simplest cameras use an internal random interlace sync generator. They usually have composite video and RF outputs. The controls include an on-off switch, target, beam and focus controls, and a mechanical focus adjustment control. The lens is usually a single 25mm unit. The camera may or may not have a built-in viewfinder.

The light reflected from the scene to be televised is focused by the lens onto the target, a disc built into the front end of the image tube. This disc is swept by a beam of electrons from an electron gun. The position of the beam is determined by the yoke, which consists of two coils of wire into which the tube is inserted. The yoke is energized by sweep voltages, which are timed by sync generator pulses. The beam follows the same path on the target that the beam of electrons in the TV set picture tube follows when the picture is recreated there. The amount of light on the target disc where the beam strikes it determines the target voltage at that moment. Thus, the target voltage varies serially as the beam covers the target area of interest, point by point, line by line.

The target voltage is fed to a video amplifier, mixed with the blanking and-or sync, and is finally available at the camera output jack. The video can also be applied to a built-in RF modulator. This produces a VHF channel signal so the camera output can be viewed on an ordinary TV set, tuned to the proper channel. The beam control mentioned earlier adjusts the beam current. When set properly it provides highlight detail in the picture. Later, we cover the full set up procedure for the camera.

The electrical focus control adjusts the beam size for the best picture resolution (detail). The mechanical focus adjustment moves the image tube back and forth with relation to the lens. It is set for best focus when the lens is focused at infinity. The lens can be a single lens on the front of the camera, a turret holding three or four lenses, or a zoom lens.

When a single fixed focus lens is used, a zoom type is normally supplied. The 25mm lens produces an image about like that seen by the eye at the same distance. A 13mm wide-angle lens can be used, which covers a greater area of the televised scene. The wide-angle lens can give the appearance of distorted perspective in some cases, especially if it is quite near the subject. Longer lenses, 50mm on up, are useful for close shots of distant objects. A 25 or 50mm type is a good choice if you are restricted to one lens. The smaller the studio, the wider the lens required to encompass the entire scene. If you are shooting in a small room you'll have to use a 12mm wide-angle type as the basic lens. The focal lengths mentioned are for the small vidicon cameras. The larger IO (image orthicon) cameras require longer focal length lenses for the same result.

If you have a turret on the camera, you can use several lenses of different focal lengths. But, be sure the view of your widest lens on the turret is not obstructed by one of the other longer lenses. The longer or greater focal length lenses are usually much longer physically than the normal or wide-angle lenses.

A turret can be moved manually at the front of the camera, or a rod controlling the turret can be run through the camera and operated by a handle at the back of the camera. Turrets are not as popular as they used to be, now that excellent zoom lenses are available.

The zoom covers focal lengths from 12 or 25mm through telephoto in one lens. To use it, you set the lens to its longest focal length and focus on the subject. It will then be in focus at any focal length selected. Of course, if the subject moves nearer or farther away, the lens will have to be refocused. The focal length can be changed by manually turning a ring on the lens or by using a remote control, usually a rod or cable between the lens and an operating knob on the back of the camera.

All but the least expensive lenses have an iris or diaphram built in to control the amount of light admitted. This control is very important to the quality of the picture. The proper setting is discussed later when we consider camera setup. Models are available for use with live studio cameras or film chain cameras, which have automatic iris control. This is a feature well worth investigating for use on your cameras.

The viewfinder resembles a small picture monitor or television set mounted on top of the camera. It has all the controls, except the channel selector, found on a TV set. These controls have no effect on the picture transmitted by the camera; they are only for the use of the camera operator to provide the picture appearance he desires. The viewfinder is supplied so that the camera operator can determine the composition and focus of the picture his camera is producing. The viewfinder can be an integral part of the camera or, in some cases, can be added to the camera later. A camera with built-in viewfinder and zoom lens is shown in Fig. 3-1.

More elaborate cameras can be remote controlled, allowing a video operator at another location to control the target, beam and electrical focus of the camera. Sometimes the iris is also controlled at that point. Cameras offering remote control usually are driven by an external sync generator. This allows fading or dissolving between two cameras, as discussed in Chapter 2, and also provides two new camera controls—video level (sometimes called contrast) and pedestal.

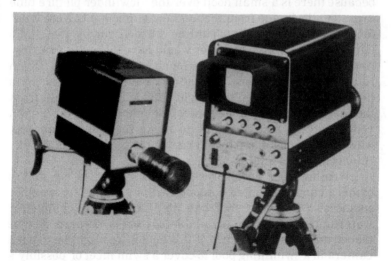

Fig. 3-1. A small vidicon camera with a viewfinder and zoom lens. (Courtesy Concord)

The video level control varies the output of the video amplifier in the camera without affecting the amplitude of the sync pulses. The pedestal control varies the black level of the video with respect to the sync. This allows setting the video black level 10 IRE units above the top of the sync pulses, as was mentioned when we discussed video earlier.

Aside from setting the levels for normal operation, the video level and pedestal controls are especially useful in setting up cameras for "supers." A good super has nice clear letters, with no extraneous white light specks in the picture. If the background around the letters being supered contains some extraneous lighter reflections, these will be seen in the super. To avoid this the pedestal can be reduced to reduce the reflections below the blanking (10 IRE) level. Then, with an increase in the video level of the white letters, a clean super will result. This is normal procedure when making a super.

Many other camera accessories are available. Intercom amplifiers and jacks are sometimes provided to allow communication with the other camera operators, the director and other personnel. Tally lights, usually found on cameras equipped with a viewfinder, light when the camera is on the air. A much smaller light is found on the back panel of the camera to remind the operator that his camera is on the air. He usually can't see the tally light on the top of the camera because there is a small hood over the viewfinder picture tube to protect the image from the ambient studio light. Were it not for this hood the image would be washed out by the studio lights and the camera operator would have difficulty focusing. Another accessory sometimes found is a filter wheel which rotates behind the lens. This allows the light to be cut by a factor of 10 or 100 when the bright outside light reflected from the scene is too great to be controlled by the iris in the lens. A yellow filter is also available to make the clouds stand out when shooting outdoors by darkening the blue color of the sky (monochrome cameras only).

Weatherproof enclosures are available also, so that a camera can be mounted outdoors with no regard for weather conditions. Carrying this one step further, a completely watertight enclosure is available so that a camera can be placed underwater. One use of this enclosure might be to put a camera in a swimming pool to cover a swim meet or possibly a water ballet. A camera in an underwater case is shown in Fig. 3-2.

The cable connecting a waterproof camera to the rest of the system can be a single coax cable, in the case of a simple locally controlled vidicon camera, or an elaborate cable with numerous conductors used to carry video to the viewfinder, video from the camera, intercom, iris remote control, and the many other control signals required to operate the camera. A standard cable for monochrome cameras with standard connector wiring is available which can be used with most cameras.

Sometimes a fully remote controlled camera is desired. Perhaps you might want to put one or two cameras in a classroom for use when instructing future teachers in their craft. A camera of this type uses a motorized zoom lens, consisting of remote zoom, iris and focus control. Also used is a remote controlled pan-tilt head, which allows the camera to be moved horizontally and vertically. The pan-tilt head is operated by a joy-stick, and the other controls are spring-loaded toggle switches. The rate of change of the zoom and other controls is adjustable. Fig. 3-3 shows one.

If more than one camera is used, they should be matched. Otherwise, if you switch between the cameras with similar

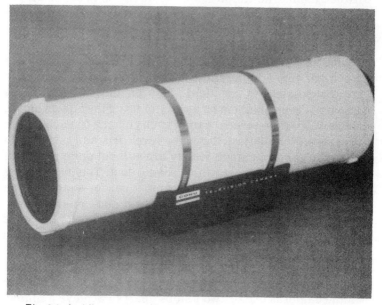

Fig. 3-2. A vidicon camera in an underwater case. (Courtesy Cohu)

Fig. 3-3. A camera with remote controlled zoom, iris, focus, pan and tilt functions.

shots there could be a noticeable difference in the gray scale between them. The simplest way to do this is to set up all the cameras (covered later in the setup procedure). Then pick one as your reference. By switching rapidly between the reference camera and one of the others you can tell if the background, clothing and fleshtones match by looking at the monitor used for comparison. Another method, which we prefer, is to get as close as possible to identical shots on the cameras to be compared. Then put them on a monitor using the effects unit to vertically split the screen, half of each camera on each side of the screen. Then adjust the video level, iris and pedestal to match the camera output. If they are still not quite matched, slightly adjust the target, which should do it. The image tubes in the cameras should be of the same type; for example, you might not be able to match a field mesh and non-field mesh type of tube. Most likely the camera tubes will be of the same type, so this shouldn't be a problem. However, matching a vidicon or Plumbicon to an IO (image orthicon) exactly is almost impossible. It is best to stick to cameras that use the same kind of tube. Incidently, sometimes you might have to run the cameras one or two f stops apart to achieve a match. Don't worry; this is normal.

Some cameras allow an external video source, other than the normal camera output to be fed to the viewfinder. This is accomplished with a toggle switch at the camera remote control position. This feature is advantageous, for example, if you want to get two cameras registered over each other for a special effect. All you need do is super one over the other and feed the result to one or both of the camera viewfinders. Then the camera operators can easily get the cameras positioned as you desire, since they can see both images. You can super the cameras on a mix bank and feed that to the preview bank, which can be wired to feed the external viewfinder inputs. Of course, you have to have a mix bank that isn't being used on the air at the moment for this procedure.

If you go for freaky effects, you can get cameras that (or modify them to) allow horizontal and vertical sweep reversal. This feature allows you to reverse the image from left to right or turn the picture upside down. This is done by reversing the yoke connections with a toggle switch (or switches) in the camera. Another effect is the negative image. This, of course, makes the blacks in the scene white and vice versa. You can also get some interesting effects if you play with the beam control, especially in color.

Another useful accessory is a frame line generator. This unit adds a horizontal white line at the top and bottom of the picture and vertical lines on either side of the picture. The position of these lines is adjustable.

To regress for a moment, the picture seen by the camera operator and the director shows the whole image seen by the camera. The viewer at home does not see about 10 percent of the picture all the way around the screen edges. This is done so the edges of the raster will not appear if the line voltage is low or sweep circuits in the home receivers are not adjusted properly. This also hides the VITS (vertical interval test signal). The VITS is discussed when we talk about video test generators. In any case, because of this overscanning, the camera operator and director must be careful that no important part of the picture extends into this unseen area. If the director's monitors and the camera viewfinders are fed these frame lines, which are adjusted to enclose the area of the picture normally seen on a TV set, it helps all concerned to remember to keep the important information within the lines. The lines are seen only on the viewfinders and the director's itors, not on the switcher output signal. The cables to the monitors and viewfinders are looped through the frame line generator, where the lines are added.

CAMERA SETUP

Setting up a camera from scratch is not at all difficult. Warm up the camera for 15 minutes with the lens capped. We have to assume you have a correctly adjusted (contrast and brightness) picture monitor or, preferably, a scope of some sort. First, set the target and beam controls about one half turn clockwise. If you have them, set the video gain about one half turn clockwise and the pedestal about one third clockwise. Set the iris on the lens to f-8 and try to focus the picture with the lens focus and electrical focus. If you can't get any focus, set the lens focus using the distance calibrations. Then adjust the mechanical focus for a reasonably sharp picture, concluding with a readjustment of all three focus adjustments for the sharpest picture. Adjust the pedestal so that the bottom of the video waveform is just above the top of the sync pulses (about 0.07 volt) on the scope. If you have only a picture monitor, adjust it until the blacks just start to spread out (cover more area). Set the pedestal control just above (clockwise) that point. Turn the target control until you get a snow (noise) free picture with good contrast. Adjust the video gain until the top of the video waveform is at 0.73 volts above the top of the sync pulses or until the whites in the picture look right (not glaring or washed out) on the picture monitor. If the whites loose detail when the target is adjusted, turn the beam control clockwise.

Next, turn the beam control full clockwise and start turning it slowly counterclockwise until the whites or top of the picture waveform start flattening out or the whites in the picture start to spread out and get a sort of shiny, pearly look. Set the beam control just clockwise from that point. Check and, if necessary, readjust the target and video gain controls. Set the lens to infinity and adjust the mechanical (back) focus for the sharpest detail in the most distant part of the scene (the farther away, the better). Then readjust the lens and electrical focus, if necessary.

Adjust the iris in the lens for the smoothest, most even change of contrast in the picture. To adjust the iris with a scope, first set the iris to its smallest opening (numerically largest f stop). Open the lens slowly, watching the scope. The whites, or top, of the waveform, will slowly come up and then stop. The vertical center portion of the waveform (the grays) will continue to increase in amplitude as the iris continues to open. Notice the f stop where the whites stop and the grays continue to rise. Set the lens about one f stop above (farther open from) that point. This is the proper iris setting for that scene. If the scene has just a few bright whites and the rest is in shadow, you will have to open the iris another stop or two.

Now the camera is setup and you should need to adjust the iris and lens focus only when you use the camera again or change scenes. If the whites in the picture seem flat when you use the camera again, check the beam setting. If the picture seems noisy, check the beam and target. The beam, target and video gain adjustments interact. If you increase the target you'll have to increase the beam setting. If you increase the target you'll probably have to lower the video gain setting. This all seems very complicated, but you'll soon make the proper adjustment without thinking about it, once you've used the camera a while.

CAMERA MOUNTING

The camera has to be equipped with some type of support, since it is too heavy to be carried by the operator, unless it is specially designed to be hand-held. The tripod is the simplest, and allows some adjustment of the desired camera height. A TV camera tripod is like the kind used for photography, but in a very heavy-duty version. If you wish to fix your camera in one position on the studio floor, perhaps on a platform, the tripod is all you will need. If you wish to move the camera about (as you probably will), you will need a "dolly," a triangular platform with three wheels. The tripod is mounted firmly on the dolly. Usually, the dolly and tripod are purchased together. Sometimes they are all one piece and cannot be separated. A folding dolly and tripod are illustrated in Fig. 3-4.

A more flexible device is called the pedestal. This is a vertical post mounted on a dolly. The post is telescopic and allows the camera to be easily raised and lowered with a crank. The center post can be equipped with counterweights to counterbalance the weight of the camera head and the

Fig. 3-4. A tripod and dolly. These units are designed to be separated and folded into small packages for transport. (Courtesy Cintel)

movable center post. This allows the camera to be smoothly raised and lowered with only slight physical effort. A large ring around the center post just below the camera provides a handle for the raising and lowering maneuver. Another smaller diameter ring below that allows the center post to be locked in any position. See Fig. 3-5.

The pedestal with the crank is steered with a handle located waist high just to the left of the crank. All three wheels can be steered simultaneously or it can be set to steer with only one wheel. The others can then turn in any direction necessary. In normal operation all three wheels are steered.

The single-wheel steering feature is supplied so that the pedestal can be oriented to position the steering handle at the rear of the camera when the camera is facing the scene to be televised. The counterbalanced pedestal is steered with the ring used to raise and lower the camera. It also has the steer one, steer three feature.

The weights in the counterbalanced pedestal are added or subtracted to exactly counterbalance the weight of the camera head. If a camera with a zoom lens is used, these weights only

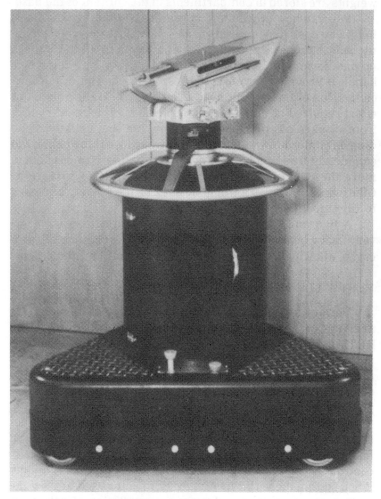

Fig. 3-5. A counterbalanced pedestal. Notice the pan and tilt head mounted at the top. (Courtesy Cintel)

have to be selected once, unless a prompter is removed or added to the camera from time to time. However, if the camera uses a lens turret and the lenses are changed, the balance weights will have to be reset to make up for the difference in weight of the various lenses. If this fact is forgotten, and you take off a long telephoto lens and replace it with a small wide-angle lens, you will find your camera shooting up in the air when the pedestal is unlocked—an un-nerving experience when it happens in the middle of a show! The extra weights are stored in compartments in the corners of the base.

You can now move the camera about on the studio floor, but if you wish to point it in various directions you need a pan-tilt head. This piece of equipment allows you to tilt the camera up and down and pan it from side to side. A long telescopic handle is provided to control the position of the camera. Knobs on the side of the pan-tilt head allow you to set the amount of effort required to move the camera. This adjustment ranges from a locked position to a completely free one. Usually, some drag on camera movement is allowed because if the camera is completely free it is hard to hold steady. It is also possible to move the camera back and forth on the head to balance it. Otherwise, the camera would be back or nose heavy and hard to tilt smoothly.

A hydraulically operated pedestal is also available for very heavy color cameras. Another unit seen occasionally is a crane, allowing the camera to be lifted far off the studio floor.

TEST PATTERNS

Various test patterns, usable in a variety of formats, are available to help adjust the camera. The most useful is the RETMA resolution chart. Among others are the multiburst pattern, the ball chart and the gray-scale chart. Let's discuss the resolution chart first, since it offers almost everything the others do. You will see from Fig. 3-6 that it is a rather complicated looking pattern.

There are five circles in the picture, one in each corner and a large one in the center. These are used to check camera linearity. When the camera's horizontal and vertical linearity controls are set properly, the circles will be properly round. When setting these the viewfinder or monitor used to view the picture must provide a linear representation. Methods of checking this are discussed when we talk about monitors.

The four black bars at the top and bottom center of the pattern are used to check the frequency response of the camera's video amplifiers. If adjustment is required you will see a black or white streak following (to the right of) the black bars. If present, they are removed by adjusting the peaking control in the camera.

The eight wedges of lines radiating from the center square indicate the resolution of the camera system. The horizontal resolution limit is determined from the vertical wedges. Where a distinct separation between the lines ends, look at the right or left of the wedge and read the number at that point. This is the horizontal resolution of the camera. The number is a measurement of resolution in television terminology. A good studio camera should read at least 550 to 600 lines. Many will reach 700 or 800 lines with ease. True, a television set is limited to 280 to 300 lines and a helical VTR to 450 at best, usually closer to 350 lines. However, the more resolution in the source video, the better the resulting picture. This does not sound very logical, but the final resolution is much crisper and cleaner than if the source is capable of only that resolution reproducible by a TV set or VTR.

The horizontal resolution in the corners can be read from the wedges in the corner circles. It will probably be less than

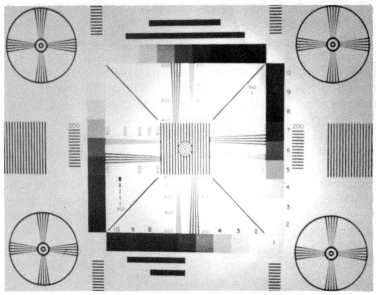

Fig. 3-6. The RETMA resolution chart.

the center resolution, but shouldn't be more than 10 to 20 percent less. In a vidicon camera, the lens and back focus, electrical focus, and, if available, alignment adjustments will affect the resolution.

When checking the resolution, be sure to notice the eight small arrow heads around the edge of the test pattern. The points must touch the four edges of the picture, or the resolution measurements will be wrong. This must be checked on an under-scanned monitor or viewfinder, where the full rectangular picture, including the corners, is visible. If an overscanned monitor or TV set is used, an exact measurement will not be possible.

If the arrows touch at the top and bottom of the picture but not at the sides or vice-versa, a width or height adjustment will be required in the camera. Before you make any adjustments be sure the camera is perpendicular to the test pattern. If one of these adjustments is made and shadows are seen in the corners of the picture, you have increased the adjustment too much. Also, if a radical loss of resolution is noted at the edges of the picture, you have also adjusted one of the scans out too far. In either case, reduce both height and width until the problem noted is corrected. Do not decrease them any more than necessary or an overall loss in resolution will result. If the height or width is adjusted, the linearity should be checked.

The square composed of steps of different shades of gray will be affected by the target and lens iris settings. The target, if set as described earlier, should be about right. The iris should be set for ten distinct steps of gray (if the picture monitor is set correctly). The scope will show a nice X in the pattern. The ten steps should be equal in height. The target can be adjusted slightly to make all the steps equal in height after the iris is used to form the X. If you set up the cameras this way, they will probably not require matching. In this case, be sure the test pattern receives the same level of illumination as the scene to be televised. This can easily be accomplished by putting the test pattern in the middle of the set, after the lighting has been adjusted.

Another test pattern, the ball chart, is used to adjust linearity. This is covered when we discuss picture monitors. The multiburst chart is a pattern that can be used for checking resolution. In use, the two patterns should form rectangles when viewed at a horizontal rate on the scope. Any loss in amplitude of the rectangles at any point along the pattern shows a loss in frequency response at that point. If the previous resolution check is satisfactory, a small loss of amplitude in the pattern is nothing to worry about, unless you are using a very expensive camera system and demand the best quality. The gray-scale pattern is the same as that used on the resolution chart; it is just easier to see without the distraction of the rest of the resolution chart information.

If you have more than one subject, or the subject is to be seen at two different positions in the set, you can sometimes use the same light as a major key in one position and a minor key in the other, and vice-versa.

LIGHTING

Lighting systems can be as simple or complicated as the studio and productions require. Lighting instruments are of four basic types; spots, scoops, spot and scoop combinations, and pattern lights. The spot throws a fairly tight beam of light, usually adjustable to some extent. The scoop throws a very wide beam of light for lighting large areas. The combination is self-explanatory. The pattern light allows the use of metal masks between the lamp and the lens to throw a light pattern on the scene. It is also used where a small spot of light is required.

At this point it is important that you understand some of the basic terms used in lighting as well as some simple lighting procedures. First of all, get a light meter that reads in footcandles. It should be an incident type of meter. Readings are made by pointing the meter towards the lights from the point where the reading is desired. For an IO (image orthicon) camera the readings should average around 100 to 125 footcandles; vidicons work better with about 200 footcandles.

The major key light is a spot and should strike the subject from about a 45 degree angle from the camera position and should read the average light level. The minor key should hit the subject at a 45 degree angle from the camera but from the opposite side. It should be about 50 footcandles less in intensity. A backlight, or light above the subject, should light the top of the head and shoulders enough to separate them from the background when seen on camera. Fill lights should be then adjusted to illuminate the rest of the set. Scoops are usually used for fill. If the fill lights have to be very bright, the key and minor key might have to be reduced somewhat to keep the overall subject lighting in the correct range. Remember, all lighting should be checked on a properly adjusted camera. This is necessary because the camera sees the lighting differently than your eye.

If you have more than one subject, or the subject is to be seen at two different positions in the set, you can sometimes use the same light as a major key in one position and a minor key in the other, and vice-versa.

The key lights model the lighting on the subject and tend to give a 3-dimensional effect. The scene could be entirely lighted by scoops and give enough light to provide a picture, but it would appear flat and not as natural as that produced when using keys. Try it and see what we mean. The key lights are produced by spotlights, which can be adjusted to cover different areas. Barn doors, or metal blinders, can be affixed to the front of the spot to shut off illumination in areas where it is not desired.

There are a few things to look out for: Keep the lights high, but not so high as to put shadows under the subject's eyes. Try to wash out most of the shadow thrown by the subject's nose with the major key. Also watch out for reflections from glasses.

A holder is provided on the front of the spot for the insertion of a piece (or pieces) of fiberglass to dim the light intensity when required.

The scoop is used for fill light to cover a wide area of the set. These can also be equipped with fiberglass sheets to dim them.

The pattern light uses metal masks with regularly perforated holes, with different size and shaped holes in different masks. They are used to project light patterns on a solid colored background. They are usually positioned to project the light on the background without illuminating the subject. The patterns should not be obtrusive, but just visible enough to provide a little interest in an otherwise uninteresting background.

Lights are usually hung from the ceiling, though floor stands are sometimes used. The instruments are provided with a "C" clamp that attaches to a grid constructed of pipes and hung from the ceiling. The pipes used to form the grid are spaced three feet or so apart. Power "pigtails" are provided for the number of lighting instruments required. A duct to carry the power to the lighting devices should be provided where necessary.

Another light mounting system provides rails that the instruments can slide around on. Pantographs are also available to use between the instrument and the grid. These pantographs are preloaded for the weight of the instrument used. They allow the instrument to be raised and lowered to the required height. The range is about six feet and allows quite a bit of flexibility in a lighting scheme. Pantographs and various lighting instruments are shown in Fig. 3-7.

Spots usually vary from 75- to 2000-watt units. An instrument that varies from 250 to 750 watts equipped with a 6-inch fresnel lens is quite popular and useful. A number of these and a few 2000-watt spots will solve most of your lighting problems. Most instruments, small and large, can be fitted with various wattage lamps. Of course, spots as large as 5000 watts are available too, but these are not too useful in a small studio. Follow spots, or stage type lights designed to put a

Fig. 3-7. A veritable forest of lights. Scoops and spots are represented. Pantographs and barn doors can also be seen.

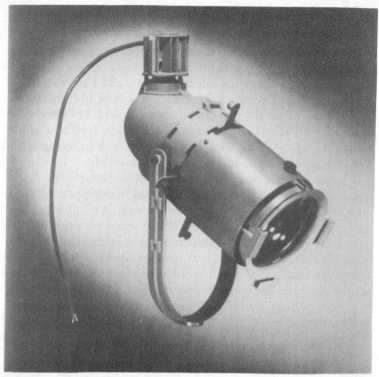

Fig. 3-8. A Leko or pattern light. (Courtesy Century)

bright spot of light around a performer, are available, too.

Flood lights, or scoops, vary from 500 to 2000 watts, depending on the instrument and the lamp fitted. The 500- and 1000-watt units are the most useful in the small studio. They will accept sheets of fiberglass to dim the intensity.

Pattern lights, or Lekos, vary from 250 to 750 watts, again depending on the instrument and the lamp. A Leko can also be used as a spotlight. In addition to the pattern masks mentioned earlier, sheets of metal that can be operated from the outside of the instrument to vary the shape of the light beam are included. Four pieces of metal are used and can be positioned to form a square, rectangle, triangle, etc. See Fig. 3-8.

You have a choice of several types of lamps to use in lighting instruments. A good choice would be quartz lamps with a nominal color temperature of 3200 degrees Kelvin. This will allow you to use the same instruments for color. Be sure the instruments you select will accept quartz lamps.

As suggested earlier, each lamp can be connected to a power pigtail, which runs to a breaker in a power panel. However, there is a better way. Each of the power pigtails can be connected to a separate plug in a lighting patch panel. Now, any of the pigtails can be energized by plugging into a jack connected to a voltage source. Patch panels can be purchased in various sizes, for small, medium and large studios. You can get almost any number of jacks and plugs you desire. One is shown in Fig. 3-9.

Fig. 3-9. A typical lighting patch panel. (Courtesy Colortran)

Another very helpful accessory is a dimmer board, which consists of a group of dimmers that allows the voltage level applied to the lamps to be varied. This, of course, changes the brightness of any desired lamp. The dimmer is usually in series with the jack supplying voltage to the plugs connected to the power pigtails. With the patch panel and dimmers, you can vary the brightness of any instrument you desire. As with the patch panel, you can purchase any number of dimmers needed. The dimmers vary in power capability. Obviously you can't use a 1000-watt dimmer with a 2000-watt instrument. The dimmers are much more practical than using fiberglass sheets to lower the light level. There are two types of dimmers, those using manually adjustable autotransformers, and electronic dimmers, usually using SCRs (silicon controlled rectifiers). The electronic units are much more compact.

More complex systems are available, allowing you to set a group of lights at various levels, then change them to other preset levels at a touch of a switch. For example, if a scene was set to represent a daylight situation, a touch of a switch could change it instantly to a night-time scene, using the same instruments but changing the light levels. Another example might be a small studio used for several programs. You could set the lighting so that it could be changed from one lighting setup to the other, simply by changing one switch position. See Fig. 3-10.

OTHER STUDIO EQUIPMENT

The studio should also be provided with surface or flush mounted floor level boxes to provide interphone, mike input and floor picture monitor jacks. As many as possible should be used to eliminate stray cords across the studio floor, which impede camera movements. The microphone inputs are paralleled together, so that a given console mike input can be used anywhere in the studio. The interphone jacks are used by the floor manager and other personnel who must communicate with the director and other personnel using the interphone line.

One or two studio picture monitors are frequently desirable in a production. These can be 17- or 23-inch video monitors, mounted on stands that allow movement about the floor. These should be good quality monitors, so they can be

Fig. 3-10. A very large dimmer system with presets. (Courtesy Century)

used to adjust the studio cameras. To eliminate problems, terminate the last video jack in the string of outlets and set all the picture monitors for bridging inputs. This will eliminate reflections. The monitors can be connected to the production switcher output or the output of a distribution switcher bank.

Chapter 4

The Film Chain

A film chain consists of a television camera especially designed to reproduce film or other projected optical images, the projectors themselves (including film, film strip and slide types), and an optical switching device to select the appropriate projector. In some cases, where only one projector is used, the optical system simply provides an image of the proper size in focus to the camera. Various combinations of projectors and cameras can be used, depending on the requirements.

You will find that projection material is very important and useful in a television system. To name just a few possibilities, slides can be used for program titles; a commercial or PSA (public service announcement) can consist of a series of slides with voice-over audio; motion picture film can be used for commercials, PSAs, or programs; a motion picture film can be used for inserts to provide part of the material in programs or commercials; slides can be used to provide titles or descriptive material to be superimposed over other video material; slides can be used as program material within a production. The list is virtually endless.

TELECINE EQUIPMENT

Telecine is a composite word for television cinema. It is a term you will hear often, and can mean the equipment used to provide projected material or it can be used to designate an area or room housing projection equipment.

Slides and 16mm film are the most popular media used in the film chain. Other film sources, though less popular, include film strips and opaque material. Included with opaque materials are graphics showing, for example, temperature along with a commercial logo, or possibly a clock instead of the temperature. Very large systems also provide equipment for the projection of 35 mm motion picture film.

Film strip projectors are usually used with educational material which is frequently provided in a 35mm format. In a production, slides are usually more convenient because they can be arranged in any order desired.

SLIDE PROJECTORS

Slides can be reproduced with a single- or dual-drum projector. The dual-drum projector provides an instantaneous change from one slide to the next. Either a movable mirror or dual projection lamps are used to perform this function. The mirror is quickly moved to reflect the image of the slide in the selected drum into the camera. The dual lamp system is simply a projector with separate lamps to project an image from either drum into the camera. Of course, only one lamp is illuminated at any one time. A popular dual-drum projector is shown in Fig. 4-1.

In a dual-drum projector, the unused drum is automatically switched to the next slide while the other drum is in use. Obviously, in the single-drum projector there is a short absence of projected light when the drum is advanced to the next slide. This short delay can be tolerated but is not desirable. To eliminate this delay, material from another source should be switched in while the projector is changing slides. This can be a pickup from a live camera or another film chain. Thus, it can be seen that two single-drum slide projectors can be used with two cameras to provide instantaneous changes between them. However, this is a rather expensive substitute for a dual-drum projector, but sometimes even two dual-drum projectors are used in this manner, so the director can see the next slide before he puts it on the air. This cuts down on the possibility of putting an upside-down or backward slide on the air. If the single-drum projector is used for titles to be superimposed over other material, the slow change cycle can be tolerated.

Where slides are used for supers, the camera pedestal or blanking control should be adjusted to darken the background of the super slide below the black level and the video gain control set to increase the white peak of the letters or image back to 100 IRE units. This gives clear, sharp letters for a key or super. The studio camera is set in the same manner when used for this purpose.

REMOTE CONTROLS

Slide projectors are available with varying degrees of local and remote control. For example, the Kodak Carousel, a single-drum projector, can be operated in a forward or reverse direction either locally or at a remote location. The lamp can be turned off locally. The RCA TP7 projector (Fig. 4-1) can be turned on and off and changed in a forward direction from a remote control point. Its local control panel provides more possibilities; it permits both forward and reverse slide changes, on-off control, and changing either drum while holding the other in the same position. Other slide projectors provide true random access, permitting instantaneous changes from any slide to any other slide in the projector

Fig. 4-1. This is a popular dual-drum 35mm projector for slides. (Courtesy RCA)

regardless of its position in the magazine. This can be a very convenient feature in some operations.

Local control centers for a film island, as they're called, can be located in various places, usually on the multiplexer itself, or a separate panel can be mounted elsewhere near the film island. The operator is thus permitted to set up the slide sequence and preview it, or cue the film projectors. The slides can be changed and the projectors operated during programming from this point if preferred. A picture monitor for each film chain camera should be within view of the film island so the operator can see what he's doing.

The remote controls for the film island, if they are to be used, should be located where they will be the most convenient for the type of operation contemplated. Some possible locations are the console or rack where the film chain camera remote control unit is installed, the distribution switcher (in a closed circuit operation), or the production and-or program switcher.

The camera remote control panel should contain all available controls, such as video level, pedestal, target, beam, electrical focus, etc. Also present should be a picture and waveform monitor or some means of setting the camera levels. The waveform monitor is just an oscilloscope, especially designed for television operation and video transmission work.

A 16mm motion picture projector is commonly used with a film chain. Some large operations use 35mm motion picture projectors as well for the higher quality offered. However, most operations will find the 16mm projector image entirely adequate. Of course, 16mm films are more easily obtained and less expensive.

ANIMATION

An animation stand will permit you to make your own animated films. Such material is used for titling, credits, and special material within a production. Basically, the animation stand consists of a 16mm camera set up over an easel. The camera control allows you to expose any number of frames desired. The camera is provided with a zoom lens so you can make the image any size you wish. The easel is a precision piece of machinery permitting movement of the photographed material in any direction for any selected distance. Such stands are very flexible in operation, permitting any desired sequence of exposures by the camera. Such equipment is quite expensive, but it is very convenient for making such films.

An alternate to film animation equipment is a VTR equipped with an editor. This is a much slower operation and it is ultimately damaging to the tape if a large number of edits are required. However, if the VTR is equipped with an electronic editor and you have a television camera you are in business, since no other equipment is required for simple animation sequences. Of course, a rigid easel stand will come in handy to hold the graphics used. You could also use slides and a film chain for the image source if the material can be obtained on slides. This is discussed in a little more detail in the next chapter on VTRs. If you use slides they must be precisely registered. The slide projector method will not, of course, permit zooms.

SPECIAL EQUIPMENT

Somewhat associated with film chains is a teletype printout pickup. Such programming can be used on a vacant CATV system channel to keep the subscribers up to date on the news. This is done simply by framing the teletype printout on a vidicon camera. You can show the actual printing process or frame only on the completed copy, arranged so it will proceed past the television camera at a smooth rate. Be sure to allow plenty of room at the edges of the picture to allow for grossly overscanned TV sets in the homes.

We mentioned opaque projectors earlier. A crude version of such a system can be useful in an ITV system, or possibly in a CATV system. This consists of a camera focused on a suitable opaque or graphic and transmitted on unused channels in an RF distribution system. The graphic can be a black card containing a printed log or schedule of the televised programs offered that day. Also, a spot on the card can be cut out and a digital clock display fitted in the hole. You could also enclose the log and clock in a circle and put a gray scale somewhere in the display. This allows TV sets in classrooms to be checked for proper linearity, brightness and contrast without feeding special test signals. An audio signal 20 db down from the normal level can also be provided to check the receiver audio circuitry. If the level is set so the 20 db down audio is barely audible, then the regular program audio will be at the proper level. The test audio can be taped music or a pickup from a background music type FM station.

MOTION PICTURE PROJECTORS

Television motion picture projectors permit the standard 24 frame per second film to be shown at a rate of 60 film frames per second to agree with the 60 fields per second TV format. This is accomplished while retaining the 24 frame per second average film speed in the following manner: One film frame is projected three times, the next twice, the next three times, etc. Thus, if you divide the 24 frames into two groups of 12 frames, one group projected three times each (3 x 12 is 36) and the other alternate frames projected twice (2 x 12 is 24) we end up with 60 frames per second (36 + 24 is 60) while retaining the normal average speed of the film. A professional television projector can be seen in Fig. 4-2.

With 16mm projectors (or any projector), it is very important that they be kept scrupulously clean. The film itself should be clean and properly lubricated. This will help keep annoying extraneous material out of the projector film gate and hold projector wear to a minimum. Any splices should be well made so they go through the gate smoothly. (More later on splicing.) The film gate itself should be cleaned before showing a film. Dirt accumulated in a gate will scratch the film. The emulsion that builds up in the gate must be removed also. This emulsion is remarkably hard and is the prime cause of film scratches. It is easily removed by using alcohol and Q tips or lint free wipes such as those used in photography. A stiff toothbrush will work well for stubborn spots. Don't ever use a metal instrument to clean the gates. Really hard to clean spots can be cleaned with a suitably shaped piece of soft wood. If a hair or other foreign object appears in the gate during projection a squirt of alcohol will usually clear the gate without having to stop the projector.

The film loops above and below the gate should be of the proper size to prevent damage to the film. The bottom loop also affects the synchronization of the sound and picture. If it is not of the proper size you will lose lip sync; this can be very annoying to watch. The proper loop size is usually designated

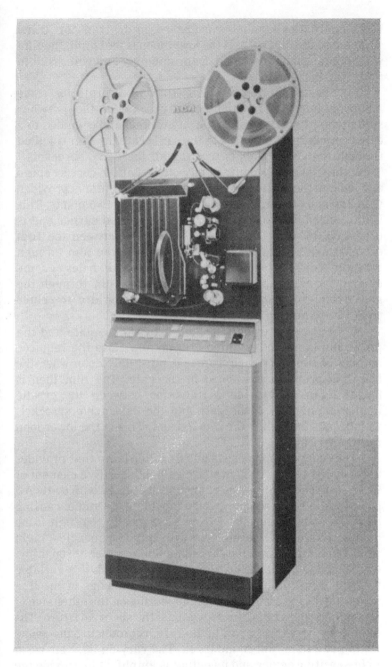

Fig. 4-2. A typical professional 16mm television film projector. (Courtesy RCA)

in the instruction manual or engraved on the transport panel adjacent to the film path. If the lower loop is too small, the film will not move past the sound drum smoothly and will result in flutter in the sound.

For those not too familiar with film projectors, the following information might clarify some of the previous statements. The film leaves the pay-off reel and is guided to a drive sprocket, which turns at the same rate the film is pulled through the gate. (This is true for all the drive sprockets.) Thus, the film is pulled off the pay-off reel at the correct speed and fed into the gate where it is projected. The film is provided with about six frames of slack between these two points. This is introduced by the upper film loop mentioned earlier and is necessary to allow the smooth transition between the first drive sprocket and the erratic movement of the film through the gate. A claw extending into the film sprocket holes or some other drive mechanism is used to pull the film through the gate, frame by frame. An adjustment is provided to center each film frame in the gate.

A "shutter" is usually located between the gate and the projection lamp and is timed to open and close the required number of times for each frame. It must only open when the frame is properly positioned in the gate. The film then is looped again to provide a transition between the erratic motion of the film in the gate and the next drive sprocket, which feeds the film to the sound drum. This is the lower loop mentioned earlier.

The sound drum consists of an exciter lamp that provides a steady light source which strikes a light-sensitive element of some sort. The light source cannot be energized with 60-Hz AC power or the line frequency would modulate the audio, causing hum. Its source must be DC or a frequency too high to be audible and too high to interfere with audio frequencies. Light from the lamp is directed through a slit and then through the optical sound track on the film. The sound track is a variable width or density strip of emulsion along the edge of the film which varies the amount of light appearing on the other side of the sound track as the film moves past the slit or aperture. The light actually varies at an audio rate, reproducing the sound contained in the film sound track. The varying light strikes a light-sensitive device and its output is amplified to provide the audio.

Some projectors are designed for magnetic sound reproduction. Instead of a light-sensitive pickup they have a head like that found on an audio tape recorder to play back sound recorded on a strip of magnetic material applied to the edge of the film and magnetized like audio tape to provide the audio signal. This method provides a better audio signal, with a wider range of frequencies possible. Some projectors can play back either type of sound, which is preferred. However, most films you will use will have optical sound.

After passing through the sound pickup point the film then runs through another drive sprocket and then through idler pulleys to the takeup reel. The takeup reel provides constant tension through a slipping clutch to keep the film taut. Not all projectors use the exact system described but most are similar. The film path in a typical projector can be seen in Fig. 4-3.

The exciter lamp mentioned earlier is subject to burn-out when least expected, so a quick method of changing the lamp is a convenient feature. If nothing else, at least you should be able to make the change without stopping the projector or interfering with the picture some other way.

It is best to rewind film by hand, rather than using the projector. Projector rewinding requires that the film be run through the drive sprockets and gate again, causing unnecessary wear on the projector and film. Besides, it is usually pretty slow. Some projectors permit you to rewind directly from reel to reel, but this risks damage to the film if the reels are not precisely aligned. Sometimes the film is not wound with the proper tension either. So it is simpler and faster to use a manual rewinding system consisting of two reel holders, one with a lever-operated brake and the other with a crank. When rewind devices are installed, they should be carefully aligned for proper operation. The film should be wound smoothly, at an even rate, for a good pack on the takeup reel. Watch that you don't clinch the film when rewinding it and that it packs smoothly. With practice only a slight application of the brake on the pay-off reel will be necessary. Be sure your reels aren't bent or you will damage the film. Again, be sure the film reels are properly aligned or all film may be damaged when rewinding.

When a film breaks it must be spliced properly, or the splice will soon part again, probably at a most inopportune

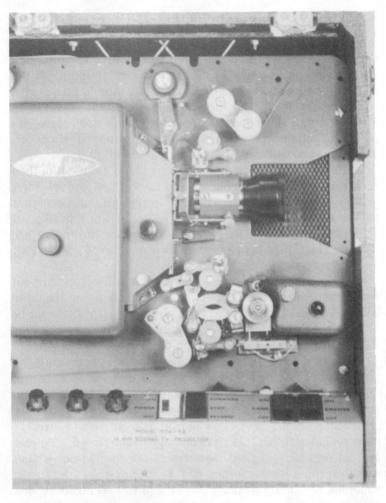

Fig. 4-3. This photo shows the path followed by a film through a typical projector. (Courtesy Kalart-Victor)

moment. Clean the film well with alcohol; this is important because it permits the splicing cement to adhere properly. Scrape off the film emulsion and binder carefully, leaving a clean smooth film base material at the end of the film. The film should be cut off square at a point allowing the cleaned bases to overlap with the sprocket holes precisely aligned. Only the bases should touch, with no overlap on unremoved emulsion. Again, be sure the film is clean at the splice point. When scraping off the emulsion and binder, be sure not to gouge the base, or the film will soon break again at that point. Clean off any excess cement after splicing, or it will get into the projector and cause no end of troubles. Be sure to let the splice dry thoroughly before snapping it or pulling on it to test its strength. Splicers are available, and are recommended. These have built-in knives and other mechanical accessories to prepare the film for splicing. The splicer also holds the film in alignment for a good clean splice. Hot splicers use gentle heat at the splice point to set the cement. Good splicing takes some practice, but the reward is the ability to make fast, easy, secure splices. Be sure to use the correct cement for the type of splice (hot or cold) made.

Film should be cleaned before it is projected. Film cleaning machines are available which do the job quickly and automatically. Film can also be cleaned by hand. Usually a liquid consisting of a wax in a suitable solvent is used to clean and lubricate the film.

MULTIPLEXERS

Multiplexers, as mentioned earlier, are used to select the proper optical input and direct it into the film chain camera. Many such systems have been developed. One system not usually seen any more had a row of projectors mounted perpendicular to a rail. A camera or cameras slid back and forth on the rail to line up with the next projector to be used. The cameras were mounted in such a manner that one could be leap-frogged over the other to line up with a projector on the other side of the camera in use. Of course, precise alignment between the projector and camera at every change was a full-time job. The systems that are popular now are a bit more sophisticated.

If more than one optical source is required in a system, the newer multiplexers use pop-up or swinging mirrors to switch the optical source. In a typical multiplexer the mirror assembly is covered to prevent stray light from entering the camera. Stray light will cause a flare effect and loss of contrast.

The mirrors in current multiplexers are designed to swing from side to side like a swinging door or move up and down in and out of the projected bright beam. The movement is accomplished by means of solenoids. The mirrors are front

Fig. 4-4. A 3-input multiplexer shown with two different types of projectors—slide and 16mm film. (Courtesy Norelco)

surfaced to avoid refraction. They are usually thin and easily scratched, so cleaning or replacement must be done carefully.

Some multiplexers provide lights on the top of the light shield to show which projector is presently in use. If a 2-camera multiplexer is used (more later), two lights are provided for each projector to show which camera is using that source. Dual-drum slide projectors such as the RCA TP7 have two lights on top of the projector to show which drum is presently in use. A tally light is mounted on each film camera to show if it is presently being used on the air.

A single-camera multiplexer usually can accept up to three projectors (one at a time). The usual arrangement is a slide projector in line with the camera and two film projectors arranged perpendicularly (one on each side) to the camera axis. An example is shown in Fig. 4-4. A typical multiplexer of this type uses two swinging mirrors, one for each side projector, positioned at a 45-degree angle to reflect either output into the camera. If both mirrors are swung out of the way, in line with the camera, the slide projector is free to beam its light into the camera. Other methods can also be used to provide the same end result.

Another type of multiplexer uses four optical sources and two cameras. A camera and slide projector are placed side by side at one side of the multiplexer, with another camera and slide projector arranged in reverse order on the opposite side of the multiplexer. A film projector perpendicular to each of the remaining two sides completes a full system. Four mirrors are used to guide any of the projector beams into either camera. Of course, the same projector cannot be used with both cameras at the same time. This type of multiplexer can be fitted with only one camera, with another added when required by additional production loads. Of course, you can also use less than four projectors, adding others later when needed. A multiplexer of this type, with the light shield removed so the mirrors can be seen, is shown in Fig. 4-5.

Another type of multiplexer uses prisms instead of a movable mirror. This allows any input to be used without doing any switching in the multiplexer. However, this type has a few disadvantages. For example, film projectors not on the air must be mechanically doused or the projection lamp turned off when rolling through the timing leader if the multiplexer is using another input. A douser is a mechanical shutter which is manually or automatically placed in front of the projector to block the light. Sometimes the douser is built into the projector light system.

SYSTEM SETUP & OPERATION

All projectors, multiplexers and cameras must be fastened securely to the floor to keep everything in focus and to reduce vibrations. All equipment must be precisely leveled as well. Believe me, this is important. Test films and slides are available to check a film chain. The 16mm film loops are used to check framing, resolution and size, and test optical tracks can be used to check the projector frequency response. The test slides are used to check the projected image size and resolution. Special slides are also used to check gray-scale

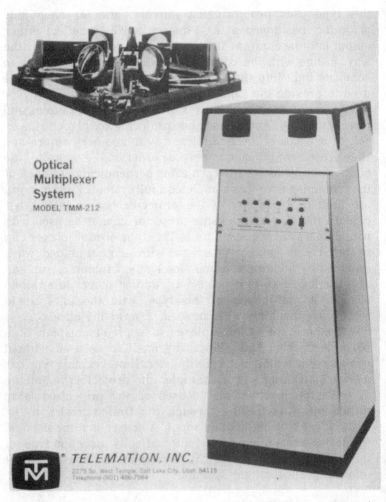

Fig. 4-5. A 4-input, 2-camera multiplexer showing the four mirrors and field lenses. The local control panel may also be seen. (Courtesy Telemation)

linearity (gamma), projector and camera iris settings, and for various tests of color film chains.

Some film chain cameras have automatic circuitry to keep the target and gain set correctly. This can be very useful for films, which can vary greatly in density, especially those made for theatre projection. The automatic controls can react faster than an operator can adjust the proper controls. This is a good feature to look for in a film camera. An internal test pulse is frequently used to make this automatic circuit set up initially. The pulse is used to make the settings, and the automatic circuitry keeps it at that level regardless of film density.

To aid in cueing 16mm film, a special leader is usually spliced on the head of the film, and commonly consists of a frame lettered, "picture start," and followed by the numbers 11 through 3 spaced a second apart. The last two seconds of leader are opaque. These numbers can be used to cue a film to a specified number of seconds before the first frame of program material. Of course, the leader is spliced to the film so as to keep the distance between the first frame of program and the number 3 exactly three seconds apart. Two common cueing points are 5 seconds and 3 seconds. The point selected depends on the time it takes the film projector to reach operating speed, the desires of the director or the delay set into the automation system.

If automation is used, the film projectors are pre-rolled a specific number of seconds before they are switched on the air. This time delay is built into the automation system. Some automation systems require manual pre-roll, others that operate by the clock automatically pre-roll the film the required number of seconds in advance. Video tapes have to be pre-rolled as well; the pre-roll time depends on the VTR and the automation system, if one is used.

In some operations a hole is punched in a few successive frames of the film, usually in the upper right-hand corner. This can be seen on the picture monitor and signals the start of a break in the film program. This break could be for a change to another roll of film or for a commercial. Usually, three cues are punched, one a warning cue, the next two for film and VTR pre-roll times. Two cues are required since the film and the VTR pre-roll times are usually different. Of course, with full automation these cues are not required, as everything is done by the clock. If a film has passed through a lot of stations, where each has punched their own cues for the times they decide to break for commercials, the film ends up with a lot of different cues. The operators have to be sharp to pick their cues out from all the others on the film. Sometimes all you can do is watch the clock and pick out the group of cues closest to the time your breaks are shown on the program log and hope you chose the right ones.

Chapter 5
Video Recorders

The video tape recorder (VTR) has been a popular piece of television equipment ever since its inception in 1956. After several other systems were previously tried and found wanting, the present system was developed. The tape used is 2 inches wide, pulled past a headwheel containing four heads, spaced 90 degrees apart. The headwheel rotates at 14,400 RPM. The heads trace slightly transverse paths across the tape, which is cupped in a concave tape guide. The tape is pulled past the headwheel at 15 inches per second (IPS). This results in a head-to-tape speed of 1,500 IPS. Each head records 17 lines of video (television lines) each time it crosses the tape. Three additional tracks are recorded along the edges of the tape—two audio tracks and a control track. The program audio is found along the top edge, the control track along the bottom edge and a second audio or cue track above that.

The early VTR (Ampex VR 1000 for example), used two or three 72-inch high by 19-inch wide racks of equipment (vacuum tube) plus the electronics and other equipment in the large console containing the tape transport. Monitors, switchers, control panels and other equipment were mounted above or along side the transport. The RCA tube-type version used a vertical transport mounted in a rack next to the other equipment racks. A third and sometimes fourth rack was used for additional optional servo and color equipment. The modern VTR, for example, the Ampex VR1200 with solid-state circuitry, is completely self contained in a unit about the same size as the tape transport console used with the VR1000. No additional equipment racks are required if the monitors are mounted over the transport. Ampex and RCA are the major manufacturers of quadrature VTRs.

Now we also have ¼ inch, ½ inch, ¾ inch, 1-inch and 2-inch (refers to the width of the tape used) helical VTRs. The helical or slant track VTR differs from the quadrature VTR previously described in the tape transport and tape path, as well as the number of heads used. Instead of the tape being pulled past a vertical headwheel containing four heads, the tape is wrapped around a drum containing one or two heads. The tape is wrapped in a spiral or helix around the drum in the case of the single-head machine, or half way around if two heads are used. The video is recorded in a long slanted track across the tape. Hence the names helical or slant track VTR. These machines also use one or two audio tracks and a control track. This format allows slow or stop motion playbacks. The quad (quadrature) VTR will not operate in a slow motion mode. For broadcast work a video disc recorder (VDR) is used for slow and stop motion. For educational or industrial applications the helical scan VTR can be used for that purpose.

The quad VTR is the machine where highest picture quality and time base stability are required. Such machines are found in all commercial and educational TV stations and production houses. Some of the large ITV systems use them, too. Many accessories are available, including a device to lock the VTR playback to the house sync generator—Intersync (Ampex) and PixLock (RCA). Other accessories to improve the time base stability for color are available. This sharply reduces any phase error (color hue shift) caused by recording error or head or guide position error during playback. Other attachments include editors, editor control, automatic frequency control of the modulator, dropout compensators and processing (proc) amplifiers (amps). See Fig. 5-1.

The editor allows electronic "splicing" of the tape. If a mistake is made during a taping session you have several choices, in addition to continuing with the taping and hoping that nobody will notice the mistake or will think you meant it that way. Of course, you can start the program over and ignore the groans and pitiful cries from the cast. Or you can stop the recording, roll the tape back to the beginning of the scene where the error occured, put the VTR back in record and go on from there. You also can continue recording after backing up the action a minute or so and mechanically splice the tape later. With the electronic editing you can back up to a convenient splice point, switch to the electronic edit mode, and when the splice point arrives in playback push the record button and continue taping from that point.

The first alternative results in an unsatisfactory program. The second leaves a lot to be desired, including the possibility of another mistake and another restart, and so on and so on. Not too good. The third is all right except that the picture will break up for several seconds at the restart point when the program is played back. This does not look very professional and shouldn't be done. The fourth method was widely used in pre-electronic editing days. However, it is hard to get a good splice in the tape, timed so it will not break up the picture. It requires a lot of time, an expensive splicer and a special fluid that allows the actual recorded tracks to be seen on the tape (which then has to be carefully removed from the tape so it will not damage the transport or heads) and a skilled splicer operator. The fact that the audio is not always recorded coincidently with the video on the tape also complicates matters. The splicing tape deteriorates the quality of the tape, too, and causes extra head wear due to excessive head tip

Fig. 5-1. A popular 2-inch quadrature VTR. (Courtesy Ampex)

penetration at the splice. The fifth method is, by far, the best and is almost always used.

QUADRATURE VTRs

The quad VTR requires a source of composite video one volt in amplitude. It does not necessarily have to be synchronous with the house sync source. However, the various segments of the program have to be synchronous with each other. For example, you cannot switch to a nonsynchronous source while recording or the resulting tape playback will roll and tear at that point. The recorder also requires a composite sync source, and, with certain accessories, also vertical and horizontal drive. The audio source should be 600 ohms, 0 dbm. The cue track can be used for many things—a second audio source for stereo, tone pulses to start and stop film projectors and change slide projectors for multimedia presentations, or tone pulses to operate an editor control unit.

A quad allows you to record both audio and video simultaneously, record either audio later with an existing video track, or video only. You cannot re-record video and keep an existing audio track unless you use an editor.

The quad VTR requires regular maintenance. The tape guide position has to be checked regularly, the video heads optimized, servo adjustments have to be checked often, filters cleaned and changed, and even the normal operating adjustments have to be set properly for each recording and playback. In short, this is not a piece of equipment just anyone can walk up to and operate without getting into trouble sooner or later; he must know what he is doing.

For example, the video head assembly is a very delicate piece of machinery and must be treated with great care. Cleanliness is imperative. Some operations run VTRs under almost "white room" conditions. Under dirty conditions you will get only about 50 hours life from a head assembly; 200 hours is quite good under clean conditions. A bad roll of tape or badly misadjusted tape guide can wear out a head in less than an hour. A picture of a quad VTR head assembly can be seen in Fig. 5-2.

The quad will run at a tape speed of 7½ or 15 IPS. Rolls of tape vary from 7200 feet down to very small rolls for commercials. Tape is manufactured by several companies. It is usually best to stick with one brand and type and adjust your machines for optimum operation with that tape. Otherwise, you will have to readjust your machine for different types of tape oxide for peak output. This is especially important when recording color.

Essentially, the VTR consists of audio, video and servo circuitry. The audio can be considered fundamentally the same as an ordinary audio tape recorder. The video circuitry, or signal system, is the portion of the VTR that records and plays back the picture information. The incoming video is applied to a modulator. This circuit changes the video to an FM radio frequency (RF) signal which is applied to the heads and recorded on the tape as it passes under each head. In playing back the tape, a switching circuit selects the head presently passing over the tape, joining each head output signal to the next in proper phase for a continous reproduction of the recorded material.

The RF frequency recorded on the tape varies according to the standard selected. The current standards are low-band, color and high-band. The early VTRs recorded monochrome

Fig. 5-2. Head assembly for an Ampex quad VTR. The tape guide is shown in the open position. Normally, it is positioned against the heads.

video as low-band. When color was used, a new standard was required, called, appropriately, color. The newer machines record both monochrome and color as high band. The difference between the standards is indicated by the frequency band used when recording the FM signal on the tape. The newer machines can record or play back all three standards, but the newer standards cannot be recorded or played back on an older machine, such as, for example, an Ampex VR100 or VR1100. This recorder can record or playback only on low band, so if one of these VTRs is included in a system, all recordings have to be made using the low-band standard for compatibility. It is possible to convert a low-band quad to high-band, but is extremely expensive. The high-band mode gives an appreciably better signal-to-noise ratio and better frequency response (better resolution).

The servo system keeps the head wheel and capstan speed in step with the sync pulses of the incoming video signal when recording, and during playback the servo locks the signal either to the power line, the vertical house sync, horizontal house sync, or both vertical and horizontal sync. With the tape playback locked to both the horizontal and output sync pulses, the VTR video output can be fed to the production switcher and treated as a synchronous source.

Accessories available for the quad include those mentioned previously. The editor, dropout compensator and proc amp are probably the most interesting. The editor was briefly discussed, but perhaps we might go into a little more detail about this useful accessory. There are two modes of editing, assemble and insert. An insert is a piece of information added within a previously recorded program or scene. An assemble is a piece of information added to the end of the program. By its nature, editing allows the additional information to be recorded without disturbing the previously recorded material. When the tape is played back the edit should look like any other switch through the production switcher. As long as the information being edited is synchronous there is no disturbance.

Before we go into editing much further we had best discuss the control track. This track contains the basic information used by the servo system to keep the playback in step with the recorded information. If it were not present, the head wheel would not be in step with the recorded information.

In the quad machine it lines up the pulses in the servo circuitry so a slight automatic adjustment of the servo can then lock the tape playback timing to the horizontal or vertical incoming sync, if either of those modes are selected. The control track looks like a rough 240-Hz sine wave as recorded on the tape. Before being used in the servo system it is smoothed into a perfect sine wave. The newer quads also have a thin negative pulse superimposed on the control track, coincident with the start of each frame of video. It is called, logically enough, the frame pulse.

In the insert edit mode, the old control track is retained on the tape as the new video is recorded. This keeps the recorder in step with the original video. When the insert is terminated the transition back to the original video will then be a smooth one, causing no disturbance in the resulting recording. The assemble mode is usually used to add material to the end of a previous recording. Since the control track was terminated at the end of the recording, the VTR will have to lay down a new control track in the assemble mode.

Normally, to initiate an edit, the VTR is placed in the selected edit mode and the servo in the auto mode (horizontal and vertical lock). Let us assume the added information is on video tape also and the edit is to be an insert. The second recorder is also placed in the auto servo mode and is fed to the editing machine. Both machines are rolled to the edit points and then rewound 10 seconds. The timer on the VTR is calibrated in minutes and seconds which makes this easy to do. We will also assume the inserted segment has been timed to be sure it will fit in the allotted space in the original recording. Both machines are then rolled in the playback mode simultaneously. When the edit point is reached, the edit machine is placed in the record mode. Actually, because of the way the editor functions, the record button has to be pushed a second before the actual edit point. This can be done by refering to the tape timer. At the end of the inserted material the stop button on the edit machine is pushed. Again, because of the way the editor operates, the machine will run on for a second or two before it actually stops. However, the outgoing edit takes place at the time the stop button is actually pressed. After making an edit, you should roll the edit machine back and be sure the edits occurred exactly where you wished them.

If the edit occured too soon, you are in trouble because you erased the tail end of the material you wanted to retain; if too late, you erased some original material after the desired edit point. Therefore, you have to be very careful of your timing.

There is a way around this problem. An editor control unit can be used. In the Ampex machine, it is called the Editec. There are various models of these units. The simplest Editec panel consists of five knobs and four pushbuttons. One knob selects normal VTR operation, insert edit mode and assemble edit mode. A fourth position allows use of a remote control panel. The second knob selects an audio-video edit or a video only edit function. These two knobs are also found on the regular editor. The other knobs and pushbuttons described are found only on the Editec. The four pushbuttons are labeled cue, pulse, inhibit and erase. A control panel can be seen in Fig. 5-3.

The cue button, with the machine in the edit mode, is pressed at the points where the edits are desired. It records a tone pulse on the cue audio track and controls the editor switches. The tape can then be rewound to a point before the first edit. The VTR is then placed in playback. The picture will switch to the recorder input when the VTR senses the first pulse and return to the playback when the second pulse is sensed. This allows you to preview the edit without actually recording it. If the ingoing or outgoing edits are not positioned to your liking, two knobs labeled start and stop allow you to advance or retard each edit point one-half second either way. If the desired change is not within this range, another pushbutton labeled erase can be pressed at the edit point, erasing the tone pulse and allowing you to press the cue button again on the next try.

When the edit points are positioned to your liking the actual editing can take place. This is accomplished by placing the machine in play at least 10 seconds before the first edit point, then pushing the record button on the transport. The VTR will not actually go into record until it senses the first cue pulse. It will then automatically switch back into play when it senses the next pulse, and so on. To check the edit the inhibit pushbutton is pressed at the edit points to supress the cue pulses. If the edit has to be redone you can be sure it will occur in exactly the same place in the original material. You also never have to worry about erasing material you wish to retain.

After the edits are made to your satisfaction, for safety's sake it is best to go back and erase the cue pulses with the erase button. The fourth button, labeled pulse, is used to switch the machine in and out of record without using the cue button and tone pulses. It is used when the exact edit point is not critical, such as when testing editor operation. When the editor is used without the Editec, a second's delay is inherent in its operation. The last knob is seldom used for normal Editec operation. Normally, it is set to the "read" position. Re-record

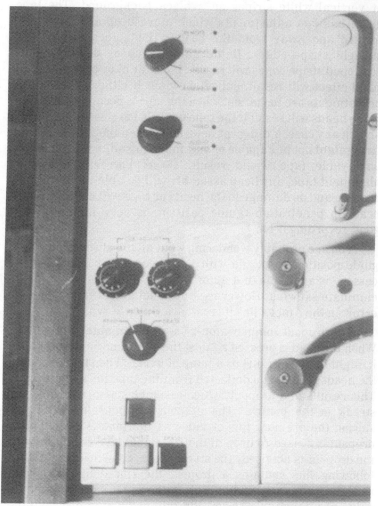

Fig. 5-3. An Editec control panel. To the right is the erase head.

is used when animating, and erase is used to erase cue pulses, just like the erase pushbutton. Other more expensive models of the Editec are available which allow frame-by-frame animation, using a great deal of automation.

We mentioned the tape guide earlier, and the importance of its adjustment. First, load an alignment tape on the VTR and place the machine in playback. It is best not to record on this tape because it is very short and very expensive. The alignment tape supplied with Ampex VTRs presents a picture of vertical white stripes on a black background. The tape guide has two adjustments which move it up and down and toward and away from the heads. The guides are set for straight stripe edges. If the guide is up or down too far, scalloped stripe edges will result; if too far in or out a venetian blind effect will result, called skewing. If either of these two adjustments are too far out when the tape is recorded, damage to the heads will result if the tape guide is too close to the heads and, in any case, a different head used when the tape is played back might not be capable of that much misadjustment and an unplayable tape would result. If set correctly with the alignment tape, any head assembly will be able to match that position and no damage to the heads or tape will occur. Proper head tip penetration (guide position) is very important for good recordings, too.

You are probably wondering what all this discussion about guide position has to do with VTR accessories. Well, an accessory is available that automatically positions the guide to eliminate skewing. However, it does not adjust for scalloping error. In the Ampex line it is called a Guide Servo.

The dropout compensator is used to eliminate "dropouts." When the head is pressed against the tape in normal operation, it might hit a bit of dirt or a lump of oxide. When this happens, the heads tends to be deflected from the tape for a short time. This results in a dropout, which resembles a horizontal white streak in the picture. The dropout compensator works by storing the previous line of video as the tape is played. If a dropout is sensed (a drop in the RF signal level) the line with the dropout is not used; the stored line is used instead. If each following line contains a dropout the stored line will keep repeating. This seldom happens, so the line is usually repeated only once. Since this occurs for only one line in one field the repeated line cannot be detected with the eye. Because dropouts are very noticeable, this is a very valuable accessory, especially if the tape is being copied. Once the dropout is recorded on another tape, the dropout compensator has no effect, since it sense no loss of RF signal.

A processing amplifier is ordinarily found in quad VTRs. It will replace damaged or missing sync pulses, clamp the signal, and allow adjustment of the video gain, pedestal settings and the sync pulse level. Clamping the signal will eliminate any low-frequency component that doesn't belong in the signal, such as 60- or 120-Hz hum. We discuss the proc amp (Fig. 5-4) further when we consider video terminal equipment.

Two-inch quads for special applications are also available; usually they are miniaturized machines. They can be used for on-the-scene tapings, political conventions, commercials, etc. Of course, they are not as versatile as the large machines, but fulfill their special functions admirably. Recordings made on the smaller units can be played back on a

Fig. 5-4. A processing amplifier used in an Ampex quad VTR. It is partly pulled out to show the printed circuit cards used in its construction.

Fig. 5-5. A miniaturized 2-inch quad VTR. The operating controls used with this recorder are on the camera. The recording format is identical to that of the large machines. (Courtesy Ampex)

large studio machine because the recording format is identical. A typical unit is shown in Fig. 5-5.

HELICAL VTRs

The 2-inch helical recorder is quite like the 1-inch type, so it is not discussed in any detail. The 1-inch helical recorder has now become very sophisticated and is widely used in ITV and CATV systems. Incidently, there are many 1-inch tape formats in use. This is a bit of a disadvantage, because if you use, for example, an Ampex VTR, your tapes cannot be played on another manufacturer's VTR. Almost every VTR uses a different format. Some formats are the same, but only because the same machine is sold under different names. We discuss the machines in general, without going into specifications of individual units because of the great number of different machines available. Sometimes you can dub to another brand of machine to change formats, but this will reduce the quality of the duplicate tape. As mentioned earlier, on a helical machine the tape is wrapped around the head drum. If a 2-head machine is used, the tape needs only to be wrapped half way around the drum. This allows much easier threading but makes the head electronics adjustments a bit more difficult. The characteristics of the two heads must be closely matched for a good picture.

The 1-inch VTR operates much like the quad; it uses a servo system which operates, in essence, like that of the quad recorder. Usually, the servo system does not allow locking the playback to the house sync, except in the most expensive machines. Editors and slow and stop motion are available on some models. Not all machines have the second audio track available, and the ability to record video and audio separately is not available on all machines. None yet have an editor control unit like the Editec. A group of 1-inch VTRs is shown in Fig. 5-6.

Some helical machines with editors require an input at the video input jack to derive sync for the servo circuitry so the playback can be synchronized to the incoming video signal. This is to assure a smooth transition between playback and record, thus ensuring a good edit. Therefore, even if you do not intend to make an edit, incoming video is required to allow the machine to play back a tape properly. If the incoming video signal is changed so that it is not synchronous with the original, a disturbance to the playback video will result while the VTR servo locks to the new sync. Thus it is well to be careful to avoid removing or changing the input to a VTR with an editor while it is playing. Some VTRs with editors will play with no video input.

Also avoid feeding noise to a VTR with an editor, such as that from a TV tuner tuned to an unused channel. The servo will be unable to lock to that type of signal and so it will be unable to function in the playback mode.

The audio microphone inputs are sometimes 250 ohms and sometimes high impedance. Check your machine so you will use the right type of mike. You can retain the almost hum free advantage of the low-impedance microphone even if your VTR has high-impedance inputs by using an external in-line matching transformer at the VTR input.

Some machines are solenoid operated, with pushbutton controls. Others use mechanical linkages to operate the tape guides and require manual operation of the fast forward and rewind modes. The solenoid-operated machines lend them-

Fig. 5-6. A group of 1-inch VTRs in use. (Courtesy IVC)

selves more easily to remote control operation. Another point to keep in mind is that some machines can be operated in only one position, usually with the transport facing up. Others can be mounted in any position, which is an advantage where rack mounting or portable operation is important.

If you have not yet decided which make and model of VTR to purchase, but have a few in mind, check with other system operators that use the type of machines in which you are interested. Ask them what they think about the model they own. Talk to several operators to get a balanced opinion. This is advisable because some machines have very poor reputations while others have excellent track records. Once you buy a machine, plus accessories for it, spare parts and a quantity of tape, you are pretty much locked into that format when you decide to expand and add more machines later. All machines produce excellent pictures when operating properly, so your main concern should be reliability. Also be sure accessories you might want later are available for your chosen model. Check on service availability, if you don't plan to do your own.

Proc amps and dropout compensators can be purchased as a part of some of the more expensive machines, or universal models can be purchased which will interface with any machine. Record time ranges from a maximum of 60 minutes in some models to 195 minutes in the IVC 900 series. Smaller reels of tape are readily available if desired.

The half-inch VTR is becoming quite popular, both as a studio recorder and for portable applications. With a few exceptions, all half-inch recorders use two video heads, which permits easy tape threading. Some also have automatic gain controls for both audio and video. This allows operation by an inexperienced person with the likelihood of a good recording. For example, a teacher could use one in the field with a minimum of instruction in its operation. The half-inch VTR is also available as a small, lightweight unit, easy to carry about. **Note.** Many machines are using ferrite heads now. They offer a long life but are quite delicate. **Always** thread a VTR while it is turned off, or with the drum switched off if that feature is available.

As with the 1-inch VTR, many tape formats are used, providing a lack of interchangeability between various models. However, a ray of hope! A new standard, EIAJ Type 1, has been devised, and most manufacturers are now making machines that use this format. This allows interchangeability between all machines using the EIAJ format.

Many accessories are also available for the half-inch VTR, as with the 1-inch machines. A half-inch VTR is pictured in Fig. 5-7. As a rule, the half-inch VTR is less expensive than the 1-inch type, but usually has a lower time base stability, horizontal resolution and a poorer signal-to-noise ratio. Your choice depends on the quality needed for your intended application. In any case, be sure to get a good demonstration of the model in which you are interested. Then decide if it will do what you want.

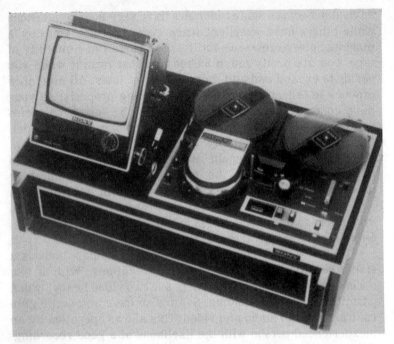

Fig. 5-7. A typical half-inch VTR. This machine has a picture monitor built into the same carrying case. (Courtesy Sony)

VIDEO CARTRIDGE RECORDER

As this is being written, the video cartridge system is pretty much up in the air, with many different systems being discussed. One such system is the Ampex Instavideo. These systems are inexpensive enough so that it might be possible to connect a playback unit of this type to each RF channel used, with a simple switch (an ordinary toggle switch would work) to select black or the VTR output. This could eliminate the distribution switcher. You could use another unit, hard wired into the studio output, to record. Patch panels could be used for dubbing, etc.

In any case, the recorder uses a small cartridge of tape, enough to last for a half hour. The recording format conforms to the EIAJ spec. No threading is required. You need only insert the cartridge and push the record or play button. It will record and play monochrome and color. Stop and slow motion are provided, too, and two audio tracks are available. Many other features are also included. The machine is very small; it weighs only 15 pounds. It is 11 x 13 x 4½ inches and will record or play for an hour on a fully charged battery. It will also operate on standard AC power with an AC power adapter.

Presently a monochrome camera is available. It will operate in conjunction with the VTR, using the camera viewfinder as a monitor when playing back a tape. It also provides remote control of the VTR and includes a 4:1 zoom lens. The complete system is shown in Fig. 5-8. Other models

Fig. 5-8. The Ampex Instavideo system. The round cartridge can be seen in the VTR. (Courtesy Ampex)

of half-inch cartridge machines are also available, together with ¼- and ¾-inch machines, too.

VIDEO DISC RECORDER

The video disc recorder (VDR) is a rather exotic piece of equipment, and you might be interested in how one works. They are frequently used on sports and variety shows.

Fig. 5-9. The disc portion of a video disc recorder. The control panel can also be seen. (Courtesy Ampex)

The disc recorder, using the Ampex HS-100 as an example, records approximately one minute of programming. It allows regular speed and slow, stop and reverse motion playbacks. The disc itself is aluminum, coated with various alloys to allow recording. Each complete revolution of the disc represents one complete television field. Two discs and four heads are used. At the discretion of the operator, one or both fields from each frame can be recorded. Each of the four heads alternately records each or every other field, depending on the mode selected. One head is erasing, one recording and two are moving to their next position on the disc as the material is being recorded. As the heads move across the disc they skip a space on their way towards the center. On the way back out to the disc perimeter they use the spaces previously left vacant. If only a single field from each frame is recorded, it has to be scanned twice during playback to form a full frame. This means the second time it is scanned it has to be delayed a half line to retain proper interlace. Any portion of the material can be selected for playback, in fact, any individual frame can be selected at will. The mechanical portion of a disc recorder is shown in Fig. 5-9.

Chapter 6
Color Equipment

Color is similar to monochrome transmission and reproduction in many respects. As a matter of fact, if you really understand monochrome you will have little difficulty in working with color.

To begin our discussion, let's go into a little detail on color TV camera operation. Then, we can use that information to discuss other uses of color in a television system. Color cameras use from one to four image tubes to form the picture. The 4-tube camera is, while usually the most expensive, also the most straightforward and easiest to understand. All other cameras derive their color signal makeup in some manner from this basic design. Before we go too much further, it might be best to define a few of the terms used with color.

Luminance is the monochrome signal without any color information. It is similar to the signal produced by a black-and-white camera. It responds only to the brightness of the various parts of the image.

Chroma or **chrominance information** is that signal which tells the color receiver or monitor what the hue and color saturation should be in each portion of the picture. This information is then mixed into the luminance information in such a manner that it has no effect on the picture displayed on a monochrome receiver but can be recovered and used by the color receiver.

Color video signal is the combination of the luminance and chroma information, with synchronizing pulses and burst added to it; these are required, of course, to make sure the lines in the television picture are spatially arranged to duplicate the original image.

Burst is a group of 8 or 9 pulses of the color subcarrier (3.58 MHz) located on the "back porch" of the horizontal blanking pulse.

Subcarrier is the frequency that provides the standard against which the chroma information phase is compared.

In all color optical systems described here, we assume there is a conventional camera lens which is used to focus the image on the image tube or tubes. We can also assume the presence of other lenses and front surfaced mirrors in the system as required by the physical layout. We discuss only the elements required for a basic understanding of each system.

The 4-tube camera obviously uses four image tubes—we'll call them vidicons in this case. The entire scene to be televised in full color is focused on the luminance tube. However, it responds only to the brightness of the various parts of the image; color does not affect the luminance tube output. The other three tubes are labeled red, blue and green. Color camera manufacturers use various schemes to separate the colors, so that only the green components of the image fall on the green tube, red components on the red tube, etc. Dichroic mirrors and-or prisms are usually used to split the incoming rays of light into the three colors. A blue dichroic mirror will, for example, reflect all the blue light and pass the rest. You can see how just two of these mirrors could be arranged to separate the three colors. Each of the four tubes treats the light received in the same way, in a manner similar to four separate monochrome TV cameras. The chroma information is then mixed (encoded) with the luminance data and the synchronizing pulses are added. The signal then can be recorded, fed to a color monitor, transmitted, etc.

The 3-tube camera system operates in a similar manner, except a luminance tube is not used. The luminance information is derived from the three color signals. A 3-tube camera is shown in Fig. 6-1.

Because of the sensitivity of the human eye to various colors, it was determined that a signal consisting of 59 percent green, 30 percent red and 11 percent blue would result in a proper luminance signal. The three signals, when superimposed on each other, produce white, and when each color is looked at separately it is equal in brightness to the other two. However, when shown as a black-and-white picture, the green will be the lightest shade of gray and the blue the darkest. The three color signals are combined as noted above and the luminance signal results. The chroma and luminance signals are then combined as in the case of a 4-tube camera.

The 2-tube camera uses a somewhat complicated method of forming the color signal. The image from the lens is split in two. One complete image is focused on the luminance tube, the other goes through a revolving filter to the red-blue tube. The filter wheel consists of two different filters arranged in strips, so that alternately red and blue images are focused on the red-blue tube. The filter wheel rotates at a precise speed, timed so that a full screen of information (one field) is produced in red, then the filter is changed and a blue field is produced. Each field is used twice, once "live" and once delayed. It is delayed precisely the time it takes for the image tube to scan one field so there is always red and blue information being fed simultaneously to the camera circuitry. It takes the camera one 60th second to scan the image on the vidicon tube so the one 60th second delay of one of the colors every other field does not affect the natural appearance of the picture.

As mentioned earlier, the luminance signal is composed of precise percentages of the various color signals. We now have two colors and the luminance information, so we need only subtract the red and blue signals from the luminance signal

Fig. 6-1. Shown is a very popular 3-tube color camera. (Courtesy Norelco)

(using an electronic matrix) and we are left with the green component of the image. Now we have all three colors and the luminance signal, which are encoded as in the other cameras.

Up until now we have used more than one tube. That means the signals from each of the tubes must be moved about by means of electronic circuitry until they are all precisely superimposed. This is called registration. A special test pattern is used for this purpose. The last camera type to be discussed does not need to be registered, since only one tube is used. While it is the least expensive (as this is being written) of the cameras discussed, it does result in a compromise of picture quality due to the method of generating the picture. For the majority of closed-circuit applications, though, it is quite acceptable.

This system uses two striped filters, fixed in position between the lens and the tube. Because of the manner in which the filters are oriented with respect to each other, red, blue and luminance signals are produced. These are all projected simultaneously on the same image tube. The red and blue pictures appear to the vidicon to be chopped into stripes. When this image is converted to an electronic signal by the image tube, it is divided into red, blue and luminance. This is done in the following manner: the filter stripes are very, very fine and are spaced by a distance equal to the stripe width. The red filter stripes are a little wider than the blue. Circuits in the camera divide the red and blue in accordance with the resulting frequency of repetition. The stripes are ignored by the luminance circuitry so we can then get a standard luminance signal. The stripes are so narrow and close together we notice no missing color information in the resulting picture. Green is derived as it is in the 2-tube camera. A 1-tube camera can be seen in Fig. 6-2.

The resolution of the picture from the 1-tube camera is limited to 240 lines in the camera being discussed. The consumer type of color receiver usually has a nominal resolution of 220 lines. Therefore, the camera works beautifully when viewed on this type of receiver. It is only on the higher resolution color picture monitors that the lower resolution becomes at all apparent.

The cameras described will work as film chain or studio cameras. They use monochrome viewfinders, when viewfinders are required, and in all other respects are mounted and handled like the monochrome cameras discussed earlier. The encoded color signal is transmitted on a single cable and treated just like monochrome with the exception of phasing, which is discussed later.

COLOR VTR

The helical color VTR works just like a monochrome VTR, with the exception of some color circuitry added to the signal system. This circuitry is used to clean up the chroma signal and allow phase adjustment. The color burst drives an oscillator to regenerate the burst signal and vary its phase so the hue of the color signal output will be within the range of the receivers' control. This adjustment is usually on the transport. The chroma is separated from the luminance signal so its level can be adjusted as well. A color VTR is seen in Fig. 6-3.

Another helical VTR color system uses special video monitors which accept a pilot signal on a separate line to provide the phase information. This is not compatible with regular monitors or receivers. An accessory is available for these recorders to provide a regular NTSC compatible color signal.

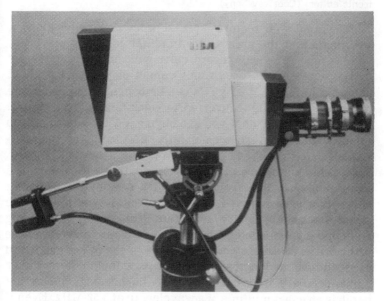

Fig. 6-2. An example of one-tube color camera. (Courtesy RCA)

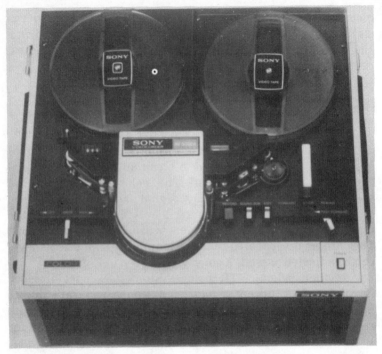

Fig. 6-3. A half-inch "colorized" VTR. This machine also has an electronic editor. (Courtesy Sony)

The quad VTR requires special accessory modules and a modification of the processing amplifier for color playback. Quite a few VTRs will record color with no modification or special accessories, though they cannot play it back without being "colorized." Most VTRs can be bought as monochrome machines and easily converted to color later, if you desire. Notice I said most, so if you are buying a VTR you think you might want to convert to color later, be sure before you purchase it.

Cable lengths are critical when dealing with color. The lengths have to be carefully adjusted so that the 3.58-MHz color subcarrier, as well as the horizontal sync pulses, are phased identically at the switchers. This permits switching from one input to another without a change in the color phase or hue. It is easiest to phase the lines at the patch panel and use identical line lengths from the patch panel to the switcher inputs. Note: Patches will destroy the phasing, unless the patch is an exact multiple of a wavelength at 3.58-MHz. Even if

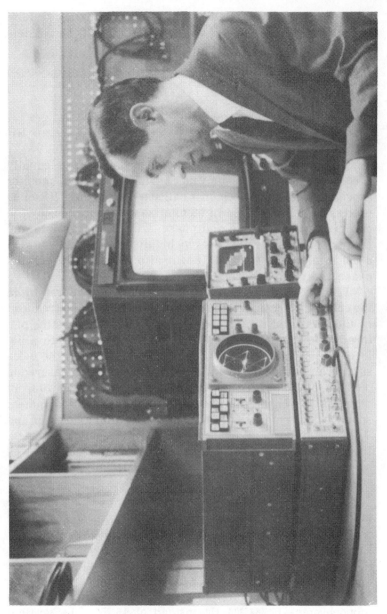

Fig. 6-4. The top unit on the left is a vectorscope. The vector presentation shows chroma and burst phase and amplitude. Under that is a color bar test generator with a built-in EIA color sync generator. To its right is a waveform monitor, with a color video picture monitor behind that. All the monitors show the color bar test signal in a different manner. (Courtesy Tektronix)

yours is not a color installation, if cable phasing is carefully adjusted, a change to color at a future date will be possible without having to rewire the system.

COLOR MONITORS

Color video monitors come in many price ranges. If necessary, a good color receiver can be used as an RF monitor.

The waveform monitors discussed earlier are equally useful for color signal analysis. Another waveform monitor especially designed for color is called a vectorscope. It can be used as a regular waveform monitor as well as to measure the chrominance amplitude and phase. We will not go into any detail about its operation but can highly recommend its use in a color television system. It will also measure differential phase and gain. See Fig. 6-4.

Essentially, differential gain is usually considered to be the color subcarrier (3.58-MHz) amplitude change introduced by the measured circuit, related in percent or db, as the picture level on which it rides is varied from blanking to white level. Differential phase is the change in phase of the color subcarrier introduced by the circuit being tested, measured in degrees, as the picture signal on which it rides is varied from blanking to white level. Broadcast quality studio equipment should have no more than 2 percent differential gain error and 2 degrees differential phase error. As previously mentioned, these measurements can be made with a vectorscope, as well as with other pieces of test equipment especially designed for such measurements.

Special lighting for color was discussed earlier when we talked about lighting in general. The main point to remember is use the right temperature lamps in the lighting instruments. The temperature should be 3200 degrees Kelvin.

The sync generator used for color must provide burst flag and color subcarrier as well as the other normal sync outputs. The horizontal sync frequency is slightly different from monochrome—15,734.264 Hz instead of 15,750. The vertical frequency is slightly different, also—59.94 Hz as opposed to 60 Hz for monochrome. These frequencies are within tolerances for monochrome transmission so the same generator can be used for either color or monochrome.

The color video test generator should provide 100 percent saturated color bars; yellow, cyan, green, magenta, red and blue in color, with additional black and white test signals. This is in line with EIA specifications for an NTSC color test signal. Usually setup (blanking level), luminance and chrona amplitudes are adjustable for various test procedures.

A black burst generator is necessary when operating with color. It is used to keep burst on the signal even after a fade to black. This is particularly important in the production switcher, because if there is no burst on the black signal the switcher may be unable to dissolve to black without fading out the burst, too. This is true only with some switchers. The burst is also required to keep the color killer turned off in television receivers. Another benefit of a black signal is that you can put some setup on the signal. This means when you fade to black it will be the same shade of gray on receivers as the black parts of the preceding picture.

Color picture monitors are required in a few locations in a color system to monitor the quality of the color picture. A color video picture monitor is shown in Fig. 6-5. You should have one connected to a good tuner to recover your signal off the cable or off the air to check the quality of the signal(s) you are sending out. You might want another at the distribution or program switcher. The director will probably want one to use as the program or preview monitor at the production switcher, too. Color picture monitors are quite expensive, so you might want to try to double up the usage on some of these to save money. Good quality color TV sets can be modified for service as color monitors in some cases, but you should have at least one good quality color monitor to use as a standard.

A color monitor can be much easier to adjust than a color TV set because all the color controls are usually on the front panel. This includes the screen, drive, and convergence controls usually hidden on a color TV set.

Special testers are available that can be pressed against the face of the picture tube over specific bars of the color test signal. The monitor is then adjusted for specific readings on a meter fastened to the back of the tester as it is placed over the different color bars. This permits all the color monitors to be set to produce virtually identical pictures. This is especially important if the color picture monitors are used in classrooms for instructional purposes. In this case there must be a locked cover over the controls so they cannot be tampered with by

Fig. 6-5. A color video picture monitor. It has no cabinet because it is designed to be rack mounted. (Courtesy Conrac)

unauthorized personnel after the monitor is set up. It is to be assumed that the program is in color because the hue and brightness of the colors are important to the material presented, such as paintings or other forms of art, chemical reactions, etc.

Of course, color systems other than the NTSC system used in the United States are available. With some systems programs can be recorded and played back on a monochrome VTR with no loss of color quality. Some systems do not require that the color saturation or hue be adjustable at the TV receiver. Of course, there are disadvantages, too, or we would be using them instead of the NTSC system. SECAM and PAL are examples of other systems in use in Europe. Of course, if you use one of these systems you are non-standard with almost all the other color systems in the USA. This obviously is a big disadvantage. You will also have trouble finding color equipment for use with SECAM or PAL, though it is available if you dig for it. Of course, the use of a color system other than the NTSC pertains only to CCTV; if you transmit a signal over the air or a CATV system the FCC requires that the color signal meet NTSC specs.

Be careful of the processing amplifiers you use with a color signal. If they are not designed for color you are likely to lose the color burst and part of the chroma signal. A color processing amplifier will probably have controls to adjust the burst amplitude, burst phase, and chroma amplitude. If the original burst is just gated through the amplifier, the burst controls will be absent. It is better to use a proc amp that regenerates the burst, because the original burst can be badly deteriorated after it is transmitted, received, demodulated, run through a few VDAs, recorded, played back, etc. If it is regenerated once or twice along the way it will be better able to perform its job when it finally gets to its destination. Of course, the burst phase must be carefully adjusted to match the original phase each time it is regenerated. This can be best done with a standard color-bar signal and a vectorscope. You can also do it by eye if you switch back and forth between the standard signal and the proc amp output signal. The vectorscope is preferred though. The vectorscope also permits setting the chroma to match the standard, too. This is harder to set by eye.

Chapter 7

RF & Video Monitors

A television picture monitor resembles a TV set; in fact, for some applications TV sets are used. There are two types of picture monitors—video and RF. See Figs. 7-1 and 7-2 for examples. The RF monitor is used in closed-circuit RF systems for viewing by the audience and to monitor the transmission in the control room.

Regular TV sets can be used or special heavy-duty RF monitors can be purchased. Some of these are available with video inputs and outputs, as well as an auxiliary audio input and output. Tuners or demodulators can also be used in conjunction with a video monitor for off-the-air reception. These provide a much better quality picture than does a TV set. One can be seen in Fig. 7-3.

VIDEO PICTURE MONITORS

The video picture monitor is widely used in TV systems. The director's position requires the presence of a monitor so he can observe the production switcher inputs and outputs. The distribution switcher should have at least one monitor with switchable inputs so the inputs can be checked if desired. This can be accomplished by using one of the distribution switcher banks to feed the monitor. If it is desirable to monitor the distribution switcher outputs, they have to loop through a separate switcher, which can feed another monitor, or a 2-input, 1-output switcher along with the input test bank. The output of a 2 x 1 switcher can then feed a monitor, allowing the use of only one monitor for checking both inputs and outputs. All these switchers should also switch the audio, so it as well as the video can be monitored. The audio should feed an amplifier and speaker in the distribution area.

As mentioned earlier, monitors should also be situated in the film chain area, close to the film camera control unit. A

Fig. 7-1. This is a 14-inch monochrome video picture monitor designed to be installed in a 19-inch rack frame. (Courtesy Conrac)

separate monitor should be provided for each film camera. The studio camera remote control area should also have a monitor for each camera. Helical VTRs should each be supplied with a monitor. It is also useful to have a program monitor, fed by the production switcher, located in the film chain-studio camera remote control area. Then the operator shading (adjusting levels, etc.) all these cameras during a production can see what the director or production switcher operator is doing. Of course, in a small system the shading position and production switcher position might be close enough together to share the same monitors, thereby considerably reducing the number required.

Most broadcast type picture monitors can be adjusted to over or underscan the picture tube. This means the monitor can be set to reproduce the picture as seen at home (overscan) or to see all four edges and corners of the picture (under-scan). Normally, the monitors in the control room are underscanned.

The video picture monitor is essentially a TV set without an RF tuner, IF section and sound portion. It usually has two video input jacks, so the monitor can be bridged across the line. This means that the output of a camera, for example, can be connected to one of the monitor input jacks, and the production switcher, etc. Next to the jacks is a termination switch which is set to terminate the line feeding the monitor if

Fig. 7-2. A small RF-video picture monitor. The tuner knob, used for RF inputs, is located on the right side, out of sight. (Courtesy Concord)

Fig. 7-3. A popular television tuner used to convert an RF television signal to a standard video signal. (Courtesy Conrac)

it is the last piece of equipment on the line. If the other jack is used, the switch is set to bridge the line (termination off).

Be careful how you set this switch, because if it is set to terminate when it should be bridging (double termination), it reduces the video level appreciably. This results in a loss of contrast and a noisy picture. If it is set to bridging when it should be terminating the line, the video level will be much too high, causing excessive contrast and a radical loss of highlight detail in the picture. If the unterminated monitor is at the end of a long line, it can also cause reflections which will foul up the monitor frequency response and also cause ghosts or multiple images in the picture. This not only affects the monitor at the end of the line, but any other equipment bridged across that line. I repeat, be sure to check these switches if you run into radical video level problems. It is very possible that the switches can be set wrong or bumped into the wrong position when connecting cables. Believe me, no end of troubles can be avoided by double checking the switch positions.

On some monitors there are jacks to allow the monitor to be driven by the house sync. Two jacks are generally used to allow bridging or termination as with the video. However, sometimes no termination switch is used, so if the monitor is at the end of the sync feed line, another method of termination must be used. This is usually a plug, like that found on the end of a video line, which contains a 75-ohm resistor. This accomplishes the same thing as the termination switch. As with video terminations, the line should be terminated at the end, otherwise the sync level on that line will be too high. Incidentally, in referring to sync, all the sync generator outputs are included, not only sync.

If the monitor is used to look at noncomposite video, external sync is usually required. However, some monitors are designed to sync on the video blanking pulses, though this is not a good practice because it tends to shift the picture a bit horizontally. When using composite video, external sync is not required, though it can be useful. For example, if the monitor is set to use external sync, it can give a quick indication when the input becomes nonsynchronous. This could be useful, for example, in a production switcher monitor. If the input became nonsynchronous, possibly if a VTR lost servo lock, the rolling picture would alert the operator to the problem. The monitor has a switch to select internal or external sync. In the internal position the monitor uses the sync pulses on the incoming video signal.

Picture monitors come in various sizes; the most usual are from 5- to 23-inch diagonal screen measurement. They are available in rack mounts or cabinets. Rack mount units are usually two 8- or 9-inch monitors mounted side by side, three 5-inch monitors mounted in a row, or one 14-inch monitor. Sometimes the 8- or 9-inch monitor is paired with a waveform monitor instead of another picture monitor. Another popular combination is two side-by-side picture monitors mounted above a single waveform monitor, which can be switched to represent either one. This might be used if you can't afford two waveform monitors for those inputs, or possibly for two film chains, if both are not likely to be used at the same time.

The cabinet mounted monitor can be used in the studio, for the monitors used adjacent to the film chains, or the monitors used in conjunction with the production switcher. Of course, one monitor type works as well as the other; the type of mounting is only for convenience. The picture monitor is one type of equipment that still can be purchased, using either solid-state or tube circuitry. The solid-state version is becoming more popular due to the need for less maintenance and because less heat is generated.

Be sure that the picture monitors are capable of **at least** as much resolution as your film and studio cameras—preferably more. This is to insure valid camera resolution measurements. If not, you can never be sure your cameras are doing what they are capable of. Also, when you look at a picture you think is soft (slightly out of focus), you should be sure it is the video source, not the monitor. Otherwise, you will have to take more time than you should to pin down the source of trouble. A good picture monitor is capable of 800 lines resolution, ample for most any video source you might be using. Of course, higher resolution monitors are available for special applications, if you need them. A picture monitor for a helical VTR will only see 350 to 400 lines at most, so a less expensive monitor could be used for that purpose if you need to cut corners. However, it is best to keep most of your monitors alike so you can switch them around in event of failure; it also keeps the required number of spare parts to a minimum and makes maintenance easier.

Another type of picture monitor you might run into is the cross pulse monitor. This allows the horizontal and vertical sync and blanking to be centered on the screen. The horizontal sync and blanking form a vertical bar in the center of the screen; the vertical interval forms a horizontal bar in the center of the screen, forming a cross, hence the name. The cross formed will show all the sync pulses with relation to each other. This cannot be seen as easily on a scope. Of course, the cross pulse monitor can be used as a regular monitor as well.

CAMERA LINEARITY ADJUSTMENTS

Earlier, we mentioned setting camera linearity while observing a pattern on the picture monitor. However, in doing so we have to accept the monitor linearity as being correct when adjusting the camera. But, we cannot be sure the picture monitor (our standard) is set correctly. Several methods are available to check this. The best approach is to purchase a sync generator (if you use one) that provides a grating or crosshatch output (Fig. 7-4). This can sometimes be added to an existing generator as an accessory, or a crosshatch generator can be added as a separate unit and driven by the sync generator. The output looks like a large number of identical size rectangles. The horizontal linearity and width controls are adjusted for equal width of the rectangles as seen on the monitor. The vertical linearity and height controls are then set for identical heights of the rectangles. The total picture area must be set to cover a 3 x 4 area; that is, the width is four units wide and the height three. The crosshatch is best brought into the switcher input as a composite video source, or made available so that it can be patched into any monitor that needs adjustment.

To go one step further, the camera can be focused on a Ball (linearity) chart. This pattern consists of a display of regularly spaced circles. A picture of this chart can be seen in Fig. 7-5. The chart also has a number of outward pointing arrow heads around its edge. These should be set to touch the edges of the camera picture. If this picture is shown on the same monitor as the crosshatch, the camera linearity can be very easily checked and, if necessary, adjusted. This is accomplished by feeding the crosshatch pattern and the camera signal to a monitor. This is done most conveniently by using the mix bank on the production switcher. If you don't use a mixer in your switcher, you can put a T connector on the back of the camera monitor, run the normal camera input into one side and a cable connected to the crosshatch into the other side of the T. This breaks all the rules about feeding properly terminated signals to a monitor, but it will give you a combination signal suitable for setting camera linearity. Use this setup only for testing because the crosshatch will be superimposed everywhere else on that camera line, too.

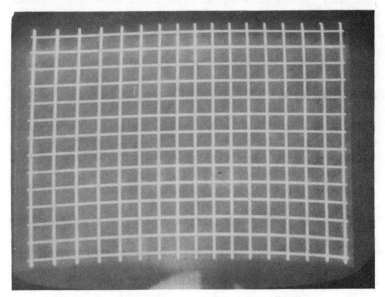

Fig. 7-4. Crosshatch pattern as seen on a video monitor.

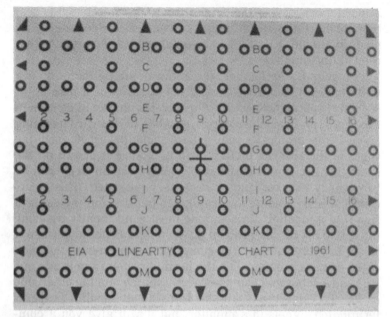

Fig. 7-5. The EIA linearity or Ball chart used to check camera linearity

First, set up the monitor correctly with the crosshatch, then add the camera signal. The crosshatch intersections should fall in the circles of the linearity chart. If they don't, try to adjust the camera linearity and size controls until they do. Of course, the camera has to be moved physically until you are as close to the right relationship as possible before adjusting the camera size and linearity controls. Ideally, the intersections should fall exactly in the center of each circle. If the best you can do is keep them somewhere in the circles, the linearity is within 1 percent of being perfect. If some of the intersections touch the lines forming the circles, you are within 2 percent. Almost all cameras will get at least this close. I am assuming the crosshatch rectangles are of the proper proportions to work with the Ball chart. This can be checked by comparing the required frequencies (plainly printed on the Ball chart) to the crosshatch frequencies as printed in its instruction manual. If you are in doubt, a letter to the crosshatch generator manufacturer should quickly answer your question. In most cases the crosshatch will be compatible. This combination of crosshatch and camera signal can be seen in Fig. 7-6.

TV PROJECTOR MONITOR

Another type of picture monitor is the television projector. This is a piece of equipment that projects a picture on a screen like a film projector. As with the film or slide projector, front or rear screen projection can be employed. The screen used is the same as that used by other optical projection apparatus. Color or monochrome projectors can be purchased. The brightness of the image projected by the less expensive units ($3,000 range) is less than that of a typical large screen film or slide projector. However, if the room is darkened enough it is quite acceptable. Depending on the equipment purchased, video and-or RF inputs can be used. A TV projector is shown in Fig. 7-7.

Some projectors lack resolution, and are best used with an image enhancer circuit. This circuit emphasizes the transitions between the shades of gray in the picture and increase the apparent resolution of the picture. This is a very useful accessory. If you plan to purchase a television projector, be sure to check its operation with and without an image enhancer, if possible. Some projectors can be purchased with

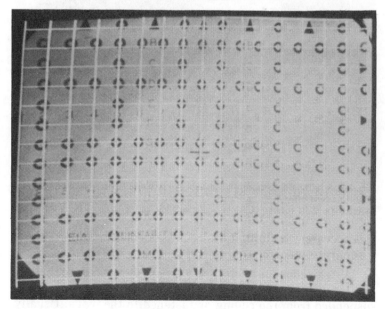

Fig. 7-6. A crosshatch pattern superimposed over the camera signal for checking linearity. All the intersections are touching the circles so the linearity is within 2 percent.

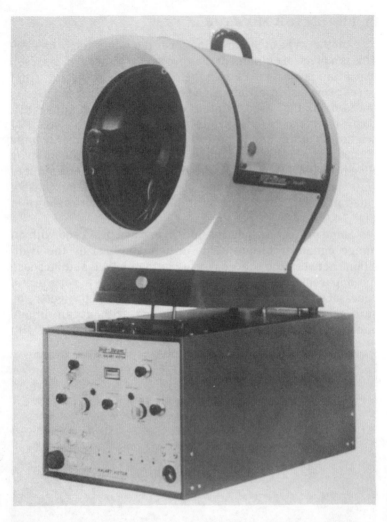

Fig. 7-7. A television projector. This model has a built-in image enhancer. It will accept only a video input. Other models also accept RF input by using a built-in tuner. (Courtesy Kalart-Victor)

this accessory built-in, if desired. It can also be added later as a separate piece of equipment.

A television projector can be used in a large hall or auditorium as opposed to conventional picture monitors suspended from the ceiling or fastened to the walls. Usually, the overall price of either method is about the same, if a half dozen or more monitors is required. When using monitors, sometimes it is difficult to find suitable mounting places where everyone has a clear view of one of them and they do not block the view of the front or rear screen used for films or slides. Another advantage of the television projector is that it is easier to maintain one projector than six or more monitors, especially if they are hung out of reach.

A television projector could be used with an RP (rear projection) screen in a studio production, too. It could provide a pre-taped animated background to a scene. Of course, matting in the special effects unit can provide the same end result, but the RP screen method leaves the special effects for other purposes. However, this requires a very directional RP screen for a sufficiently bright image and the lighting must be carefully planned to keep stray illumination off the screen to avoid washing out the image.

WAVEFORM MONITOR

The waveform monitor so frequently mentioned is a special oscilloscope for television control room monitoring. It is also used to monitor video in very large distribution systems. A regular scope can be used as well where required, but if only one is used it has to be specially connected everytime it is needed. Of course, waveform monitors are expensive so it is up to you to decide where convenience outweighs cost. Ideally, waveform monitors should be used to monitor studio and film camera outputs and the output of the production switcher.

Most waveform monitors have two inputs so they can monitor two video lines, as stated earlier. If necessary, one waveform monitor can be used for all the signals in a given area by using a small distribution switcher to feed the monitor. A possible disadvantage to this scheme is that an inexpensive passive video switcher, which might seem ideally suited for this use, has some unfavorable characteristics. Most passive switchers are terminating types, so they have to be at the end of all the input lines and can be the only terminated input device on each line. When this type of switcher is operated, transients on the video line being switched on and off are evident in every other piece of equipment connected to that switched video input line. Use of an active switcher would solve some of those problems, but would introduce some new ones. It also is more expensive; however, it will eliminate the transient problem and will bridge the line instead of requiring termination. However, it contains an active amplifier, the gain of which has to be checked frequently to be sure it remains at unity to provide true measurement of the signal amplitude. You also have to be sure it does not add distortion to the signal. A third (and best) choice is to use a passive bridging switcher. It has none of the disadvantages of the preceding switchers, but only needs to be located very close to the waveform monitor. Of course, the waveform monitor must have a bridging input, which should be no problem.

The use of a waveform monitor is not really necessary, because, as noted before, a regular quality oscilloscope can be used. However, the waveform monitor is more convenient to physically mount in an equipment rack or console and, since it is designed expressly for that service, it is much easier to operate. It also has a few features not normally available on a scope, such as specially marked graticules, filters to vary the frequency response, and the ability to select and look at a single line in the picture. In the preceding description, we were referring to the Tektronix 529 series, but other manufacturers offer waveform monitors similar to the Model 529. A somewhat less expensive Tektronix monitor is also available, the 528.

It should be informative to examine a typical waveform monitor and briefly outline the uses of its controls. We will use the previously mentioned Tektronix line of monitors, this time the RM 529, for our example. This is a 5¼-inch high unit of the proper width to mount in a 19-inch equipment rack frame. It is mounted on slides, so it can easily be removed from the rack for maintenance. We will divide the front panel into three sections, left or vertical, center or beam, and right or horizontal. See Fig. 7-8.

On the vertical or left side, notice the input switch, with positions designated Cal, A, B, and A-B. The Cal position puts a square-wave calibration signal on the CRT (cathode ray tube) which allows a check of the vertical gain of the monitor. The amplitude of this signal is selected by a switch at the lower right of this panel. There, a 1-volt, 0.714 volt, or external calibration signal can be selected. If this calibration signal does not show exactly the right amplitude on the graticule, the gain control (screwdriver adjustment above the calibrate switch) should be adjusted to set the gain exactly. The external calibration input is used where a common calibration signal, provided by an external standard, is used to set all the waveform monitors to a common standard. Normally, the internal calibration signal is suitably accurate. The A and B positions select either of two video inputs. The A-B position gives a differential signal, resulting from the mixture of both inputs. The position control moves the waveform up and down on the display tube (CRT). This allows setting the waveform to the proper point on the graticule for accurate measurements.

The response switch connects various filters to allow numerous tests to be made using the scope. Normally, it is operated in the flat position. To the right of this switch is an amplitude control which adjusts the vertical size of the display on the CRT. The large part of the knob adjusts this amplitude in steps; the small part of the knob allows a continually variable change within the limit of the step. Normally, the large knob is set to 1.0 and the small to a fully clockwise position, which lights the calibrate light above and to the right of the knobs. Below the response knob is the DC restorer switch. It should be operated in the on position. The switch in the off position allows the waveform to float up and down on the CRT as the average video level changes. This is not desirable. So, the switch should be in the on position and the sync switch in the right or horizontal section should be in the external position. If not, the clamp circuits will not work and thereby will not allow the DC restorer to operate.

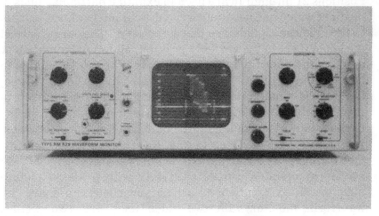

Fig. 7-8. A waveform monitor showing all the controls described in the text. (Courtesy Tektronix)

The center section contains the power on-off switch, the CRT and a few CRT controls. Various graticules are available, depending on the use to which the monitor is put. Usually, the graticule used is the composite with 140 IRE unit rulings. To the right of the CRT is the focus (which sharpens the waveform image), intensity (sets intensity of the waveform image) and scale illumination (sets brightness of the graticule edgelighting).

The horizontal or right section has a position control which adjusts the horizontal position of the image on the CRT. The Mag (magnification) switch allows expansion of the signal for detailed analysis of the waveform. Normally, it is set at X1 (no magnification). Below this is the field switch which selects either the even or the odd field to be placed to the left on the display. To the right of this is the sync switch; it selects either internal or external sync to trigger the monitor sweep. As previously mentioned, external sync is used ordinarily, permitting proper operation of the DC restorer circuit. The display knob operates in conjunction with the line selector below it. Normally, the display switch is set to the 2 Field or 2 Line to monitor the horizontal or vertical aspect of the signal. The 0.125 H per cm and 0.25 H per cm positions allow expansion of the sweep for special requirements.

A CCW rotation of the display switch from the 2 Field position brings the line selector into operation. The line selector permits viewing any selected line in the fields. Positions 21 through 16 are used to examine particular lines in the field which are used to carry vertical interval test signals (VITS). These signals are discussed later. The large part of the knob, if set to the variable position, allows examination of any line in either field. The small part of the knob rotates to permit this selection. To determine which line is being viewed, the picture monitor used in conjunction with the waveform monitor must be connected to the video output jack on the back of the waveform monitor. The single line selected is then brightened with respect to the rest of the lines forming the picture. A disadvantage of this feature is that the waveform monitor can be used only with a single picture monitor. If, as mentioned earlier, you choose to share the waveform monitor with two video lines and two picture monitors, this feature must be sacrificed. You can still use the single line feature for VITS examination, but if you use the variable mode, you have no way of knowing which line you are viewing.

The rear panel of the waveform monitor contains jacks for a bridging input to both video inputs and the sync input. The external calibration signal input is also a bridging input. If a signal line is ended at the waveform monitor, a termination plug should be inserted into the unused jack to terminate the line. Termination switches are not usually provided. The other jack is the video output discussed earlier. It should be terminated if not used. In general, unused outputs, video, pulse or audio, should always be terminated as a matter of course. This refers to all types of equipment, not only monitors. Waveform

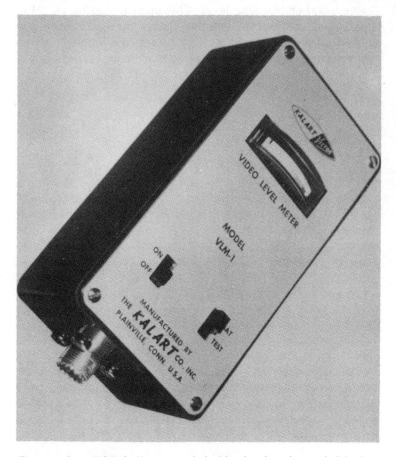

Fig. 7-9. A portable battery-operated video level meter and detector. (Courtesy Kalart-Victor)

monitors other than the Tektronix 529 have essentially the same controls, though their positions will be different.

A piece of equipment is available that will superimpose a waveform of the signal over the picture monitor. This eliminates a need for a separate waveform monitor but gives a similar end result. This is not so elegant as using a separate waveform monitor or quite as versatile, but it is a usable alternative and will save you some money.

Yet another alternative is the use of a detector and meter. This allows you to set your signal to the proper amplitude but gives you no idea of the waveshape of the signal. If you are concerned only with levels at that point, it will do the job admirably for relatively few dollars. The meter will not indicate the pedestal or blanking level. This must be set by visual indication on the picture monitor. These meters can be purchased as separate units or sometimes are found built into camera controls. An example of a separate unit is seen in Fig. 7-9.

Chapter 8
Video Terminal Equipment

This is a sort of catch-all chapter covering video distribution amplifiers, clamping amplifiers, balanced-to-unbalanced video amplifiers, patch panels and processing amplifiers.

VIDEO DISTRIBUTION AMPLIFIERS

.... A video distribution amplifier (VDA) is a single-input multiple-output device. It is used where a single video output must feed several pieces of equipment located at physically far separated points or where several pieces of equipment have terminating inputs. See Fig. 8-1. A video distribution amplifier usually has a bridging input and four to six outputs. It should have a common gain control for all the outputs as well as a test point for the inputs and outputs.

Some VDAs are self-contained units, with a built-in power supply or a group of VDAs can be plugged into a common rack-mounted enclosure with one or more power supplies. Sometimes two power supplies are used to feed a group of DAs. In such cases, usually either power supply is capable of doing so alone. An automatic circuit cuts out one of the supplies if it fails, allowing the other to carry the full load. A warning light usually is illuminated, warning the operator which supply failed. It can then be repaired or replaced, while the VDAs continue operating. This is a nice feature, but its

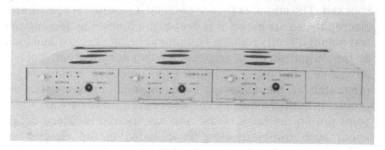

Fig. 8-1. Three video DAs mounted in a rack frame. Notice the test points available. (Courtesy Cohu)

added expense is justified only if a failure would have serious enough consequences to warrant it. Unused VDA outputs should be provided with terminators. Some VDAs come with differential input circuitry which is designed to eliminate any ground loops causing hum in the signal. This is a valuable feature for a DA at the end of a very long line.

VIDEO TEST GENERATOR

A video test generator provides various test signals for use in testing a system or individual pieces of equipment. Several signals can be provided; the most common are multiburst, stairstep, color bars, sine-squared and window.

The multiburst signal consists of six groups or bursts of sine-wave signal and one broad white pulse. From left to right, there is the afore mentioned white pulse, a pulse burst at 0.5 MHz, the next at 1.5 MHz, then 2.0 MHz, 3.0 MHz, 3.58 MHz and finally 4.2 MHz. This will probably be clearer when you look at Fig. 8-2. The frequencies may vary with the model of generator, but those mentioned are typical. This is a most useful test signal to check the frequency response of a piece of equipment. The pulse is the reference amplitude. Ideally, the bursts should all be the same height when viewed at the horizontal rate on a scope or waveform monitor. If the bursts increase in amplitude as the frequencies increase, the high-frequency response of the equipment under test is too great; conversely, if the bursts decrease in amplitude as the frequency increases, the high-frequency response leaves something to be desired.

To use this or any one of the test signals, the test signal is applied to the input of the piece of equipment or system being checked. The output is then viewed on a monitor scope. If any discrepancies are noted, it is best to pin it down by feeding the test signal directly into the equipment being checked and look at the output directly across a 75-ohm termination. This is advisable, because sometimes a defective cable or a mismatch at the following piece of equipment can cause problems in the frequency response. Terminations with a test point pin on the back of the unit are available. More detail about such troubleshooting is included in the chapter on maintenance. Incidently, if the previous tests show a discrepancy, be sure to check the input signal right at the input to the equipment in question by terminating that line after disconnecting it from the equipment being tested, just to be sure. When chasing troubles, never trust anything, not even a piece of cable.

Normally, every piece of equipment, except a VTR or equalizing amplifier, should show a flat frequency response when testing with multiburst. A VTR will show about 20 IRE units roll-off (decrease) of the last burst if it is a quad type, and more if it is a helical VTR. This is to be expected. The equalizing amplifier can show a gain or loss of high-frequency response, depending on how it is adjusted.

The high-frequency gain or loss in the equalizing amplifier is normal, this being what it is designed to do. It is used to make up for the loss of high frequencies over a very long video line (several hundred feet or more). It accomplishes this by increasing the high-frequency response enough to make up for

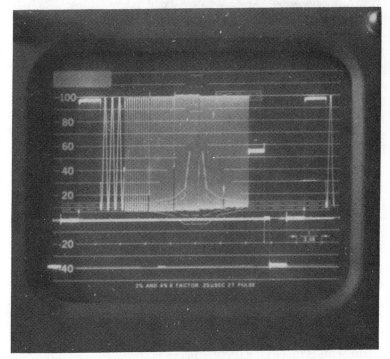

Fig. 8-2. Horizontal representation of the multiburst test signal. You can see the reference pulse on the left and the six bursts, each at a different frequency.

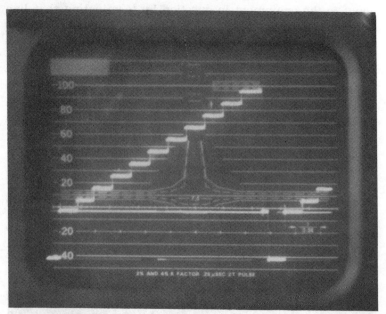

Fig. 8-3. Horizontal representation of the stairstep test signal. As you can see, this is the 10-step version.

the losses in the cable. The amount of equalization is adjustable, using controls or switches in the amplifier.

The test signal in Fig. 8-3 is called a stairstep. This can consist of three, five, or ten equally spaced steps, from blanking to peak white. Normally, ten steps are used. After passing through the amplifier or other equipment being tested, the amplitude of the steps should still be equal. If the step amplitudes are not equal, the equipment gain is varying with the average picture level (APL)—a defect which should be corrected. The stairstep signal is also useful for setting the brightness and contrast controls on a picture monitor. In this case, the brightness and contrast are set for ten distinct shades of gray. The brightest bar should be set for the desired maximum picture brightness and the second darkest just above the blackest possible setting of the monitor. Each step of the stairstep signal can be modulated with a 3.58-MHz (color subcarrier) signal for measurements of differential phase and gain. This is discussed further in the maintenance chapter.

The sine-squared and window test signal is also useful. This is actually two test signals in one. The sine-squared pulse is a single, extremely steep-sided pyramid shaped pulse. It is used to check for amplitude versus frequency response, transient response and envelope delay and phase. Proper high-frequency amplitude characteristics should be seen as a 100 IRE unit symmetrical pulse on a waveform monitor displaying the horizontal rate. Improper transient response is represented by overshoot on the positive or negative transition of the pulse. Ringing, or a ripple following the pulse, means increasing delay with increasing frequency; if the ringing precedes the pulse, decreasing delay with increasing frequency is indicated. Properly operating equipment will show no distortion of the pulse. This signal is shown in Fig. 8-4.

On a picture monitor the sine-squared signal looks like a single, sharp thin vertical line preceding a large white square (the window signal). The square should be sharp and clear, evenly illuminated with no ringing or streaking to the right of the signal. On the scope, the sides should be vertical and the top horizontal. No disturbance following the window pulse should be evident. Other test signals are also available, too, but these mentioned so far are by far the most used.

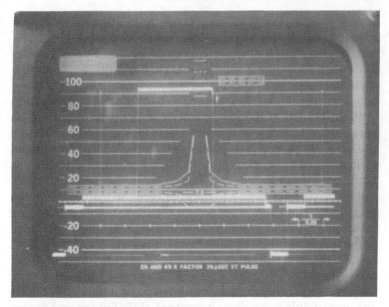

Fig. 8-4. Both the sine-squared pulse and the window signals are seen in this photograph. This is a normal horizontal representation and the sine-squared pulse is not expanded to show the leading and trailing edge disturbances mentioned in the text.

A sequential switcher provides an interesting method of using these various signals. If the test generator is a modular unit, with each signal always available, the outputs can be connected to the inputs of a sequential switcher. This type of switcher automatically switches from input to input at a predetermined rate, usually adjustable. If more inputs are available than you need, or you wish to bypass one or more of them, you can do so by a simple adjustment of the switcher. You can thereby vary the test output at the predetermined rate. This allows using all the test signals available without the necessity of having to go to the test generator and manually change the output to perform the various tests. If desired, you can usually stop the switching sequence at any desired input. Such switchers vary, with different features available on different models. Most are bridging switchers.

Another use of such a switcher might be to provide test signals, correct time and program information on the unused channels in an ITV system. You could also provide music 10 db or so down from the regular audio level.

VERTICAL INTERVAL TEST SIGNAL

VITS is a term you hear frequently. It means vertical interval test signal. Various test signals can be inserted into one or more lines in the vertical interval. The FCC says lines 18 through 20 inclusive may be used to carry VITS. Normally, these lines are vertical blanking. On a TV set they are not seen because the picture is overscanned to hide the horizontal and vertical blanking. If the vertical hold is misadjusted so the vertical interval can be viewed, the VITS will be seen between the vertical serrations and the beginning of the picture information on network programs. If a waveform monitor has the capability to look at one line (such as the Tektronix 529), the VITS can be viewed and the quality of the transmission can be judged. The VITS is added to the selected signal by looping it through a VITS keyer. In some cases, this keyer is provided in the test generator; in other cases, it is an accessory.

The clamp amplifier is used to eliminate low-frequency modulation of the signal. This looks like a 60-Hz disturbance in the baseline of the video signal, viewed at a vertical rate. In the picture this looks like alternate slightly lighter and darker broad horizontal bars in the picture. These can be caused by a difference in grounds between each end of the cable. Since the cable shield carries the video signal too, it will add a 60-Hz signal (the AC potential difference between the ground points developed across the reactance of the cable shield and ground and connector connections) to the video if it is not grounded at the same potential at each end. The clamp amplifier will remove this unwanted signal by clamping or returning each sync pulse tip to the same DC potential. The 60-Hz hum on the signal changes the video so slightly from one horizontal sync pulse to the next that it is no longer noticeable once the sync tips are clamped. Usually, only a long line will produce this kind of disturbance, for example, from one building to another. Each building using a different ground system (ground rod or whatever), might provide a potential difference from one end of the cable shield to the other. Within the system, poor grounds can be caused by poor connections between the ground buses or insufficient size of the ground conductors. This is covered in more detail in the chapter covering installation.

Another clamping method returns the horizontal blanking to a predetermined DC potential. This accomplishes the same thing as sync tip clamping, but the video clamping level will not vary if the sync level changes. Usually, blanking clamping samples the pulse "front porch" level.

PROCESSING AMPLIFIER

A processing amplifier clamps the signal, as does the clamp amplifier. It also includes a sync generator so the entire sync pedestal (sync pulses and blanking) can be replaced in phase with the original information. Most will also regenerate the original sync pulses if that is desired. Either method can be useful when using an off-the-air signal because the sync pulses are then usually quite distorted and noisy. An automatic gain control (AGC) circuit is sometimes provided, which holds the video gain to a predetermined level. To allow fades to black, the AGC circuit often relies on the sync level to set the video gain. Therefore, it is important to keep the sync level constant when this circuit is used. Proc amps without AGC circuits frequently contain a white clip circuit. This circuit cuts off or limits any video over a certain level, usually 110 IRE units, to prevent excessive deviation of a VTR modulator or overmodulating a TV transmitter. The result of overdeviation or overmodulation can cause black streaks in

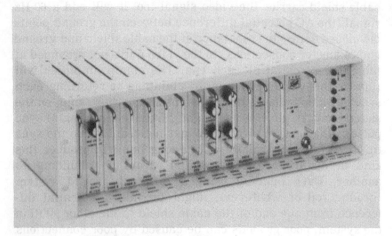
Fig. 8-5. A video processing amplifier using modular construction. (Courtesy Ball Bros.)

the picture or a loud irritating buzz in the accompanying audio. A processing amplifier is shown in Fig. 8-5.

Some helical VTRs have a gap of several lines in some portion of the video where the head leaves one edge of the tape and re-enters at the next edge. This interval varies in width and position in different recorders. In any case, it is advisable to replace the missing sync pulses in the signal. The proc amp will do this very nicely. Some recorders drop the lines at the bottom of the picture, just before the vertical interval. Some TV sets mistake this gap for vertical sync, and the result is vertical jitter in the picture. The proc amp eliminates this problem. You can see, a proc amp is a useful piece of equipment to have around. There will be many times when you will use it, especially when dubbing between two helical VTRs or recording off the air.

If you are running video over a very long distance, you can run into interference problems, as discussed earlier. However, there is another solution. Balanced video cable, which has two conductors in a shield, is available. The impedance of this cable is 124 ohms, as opposed to 75 ohms for a conventional coax line. It is balanced, and so is nearly immune to outside interference and is not affected by a potential difference between the ground points since the shield is not part of the video circuit. Amplifiers are available to convert from one impedance to the other. Usually, these amplifiers contain equalizing circuitry, or are used in conjunction with a separate equalizing amplifier. The equalizer can be located at either end of the line, sometimes at both ends when a great deal of equalization is required. The 124-ohm cable is relatively inexpensive, as are the amplifiers, when excellent video quality is necessary.

PATCH PANELS

Patch panels are another useful accessory in the video system. It allows the circuits routed through the patch panel to be interrupted and rerouted if a nonstandard system configuration is desired or a piece of equipment fails. The equipment outputs are connected to the top rows of the patch panels; the bottom rows feed inputs. The patch panel consists of two horizontal rows of patch cord jacks, and the cables are connected to cable connector jacks on the back of each of the patch cord jacks. Each vertical pair is normally connected together, either internally (normaled through) or via a looping plug inserted from the front of the panel. Thus, each vertical pair should be associated; that is, the top output jack is normally connected to the lower input. The system is thereby set up to operate in its designed mode with no patches or with just looping plugs. Patch cords should be used only for a nonstandard requirement. If an input (lower jack) is wired for a spare piece of equipment, the upper jack is left unused. If a spare output is connected to an upper jack, the line is terminated with a termination plug on the back of the unused lower jack or a patch plug termination is inserted into the patch panel from the front. Front terminations are almost always used with panels requiring looping plugs. Unused inputs need not be terminated.

It is not a good practice to pair unassociated inputs and outputs in a looping type panel. This is not technically unfeasible and it does help use up unused jacks (which are expensive), but makes the panel very confusing to use. Of course, if normaled-through pairs are used, the above practice is unlikely to be considered. Some jacks have an extra jack alongside each output, with no jack below it. These jacks are found where looping plugs are used. They permit a scope to be bridged across the circuit so the signal can be sampled there without removing the looping plug, which would break the

Fig. 8-6. A video patch panel. This unit is enclosed for shielding. Normally, this is not necessary. (Courtesy ADC)

circuit. A double patch panel can be seen in Fig. 8-6. A patch cord for this type of panel is shown in Fig. 8-7 and a looping plug in Fig. 8-8.

Exactly the same hardware and methods are used for sync patching systems. For example, you might want to route all the sync generator outputs, the pulse DA inputs and outputs, and the system pulse lines to the sync patch panels. If a sync generator changeover switch is used, you will probably want to route just the two sets of sync changeover switch outputs to the patch panels, while connecting the sync generator outputs directly to the changeover switch inputs.

The sync generator remote or external input is usually connected to the video patch panel, since video is the type of signal patched into that input. The crosshatch or bar-dot output is also connected to the video patch panel, since it is essentially a composite video signal and is treated as such.

If sync encoders and decoders are used, they are also connected to the sync patch panel. This allows any encoder to be patched into any decoder desired. Also, any encoder can be patched into any sync generator, permitting that encoder and any decoder together with its driven equipment to be locked to any source. This can be very convenient if you only want to lock up part of the equipment to another standard while keeping the remainder on the regular house standard. Having all the encoders, decoders and sync generator outputs connected to a patch panel makes such switching a simple operation.

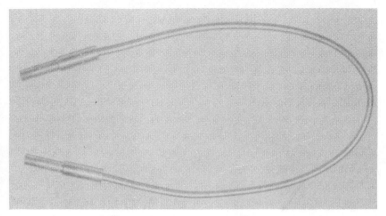

Fig. 8-7. A patch cord used with video and sync patch panels.

Ideally, all circuits, both video and pulse, can be routed through the patch panels for maximum flexibility. Patch panels are very expensive, so this is not usually a very feasible plan. Therefore, you must carefully choose the points in the system where the patches will do you the most good. Some possibilities are switcher inputs and outputs. Spare equipment that is mounted in the racks should also be brought to the patch panel, assuming the equipment like it is also brought to the patch panel. Other possibilities were mentioned earlier. If you notice where the patches are placed in the systems in the last chapter, perhaps you might get an idea or two. In any case, you know what you want to do with your system better than anyone else, so you are probably the best judge of where the patches should go. Be sure to use insulated jack panels to avoid ground loops between equipment due to unwanted cable shield cross connections through the jack field panels.

Fig. 8-8. A video looping plug.

Chapter 9
Distribution Systems

There are two basic types of distribution systems to consider—video and RF. Video is usually distributed with accompanying audio.

CABLE

Two types of cable can be used, coaxial (single-conductor in a shield) and balanced video (two conductors is a shield). For runs of any length, the balanced video line is the better choice. Both conductors are isolated from ground and wound together in such a manner that they tend to cancel any interfering signal impressed upon them. The ground is used as a shield to further protect the video signal from extraneous noise. With coax cable the shield protects against interference and serves as half of the signal carrying circuit. If the ground potential at each end of the coax cable is different, a 60-Hz signal (power line) will be impressed on the shield and thence on the video signal. This appears as darker and lighter broad horizontal bars in the picture. In bad cases the picture will show severe distortion and bending.

For short runs of coax cable (several hundred feet or less inside a single building), no problem usually is evident in the picture, although some hum might be noticeable in a field rate oscilloscope display of the signal at the monitor. Very long runs, even to another building, can be used if the monitor is insulated from ground or has a differential input circuit. The shield should be jacketed so it is not grounded anywhere along its path. In any case, shielded cable should always be jacketed, whether carrying audio, video or RF.

For long runs of cable, equalizing amplifiers are used to keep high-frequency signals at the proper level. Pre-equalization (that is, equalization at the sending end) is sometimes prefered. This lessens the liklihood that any high-frequency noise, picked up by the cable when the video high frequencies are boosted, will be amplified. However, this is not usually critical unless the noise is quite bad. It is easier to adjust the equalization by checking a multiburst signal with a scope directly at the output of the equalizing amplifier located at the receiving point.

The differential input mentioned earlier permits you to isolate the coax cable shield at the receiving point. This will eliminate the difference in ground potential between the sending and receiving points. Amplifiers with this type of input are frequently used at the ends of long video lines. They also frequently include equalization circuitry as well. The amplifiers can be set for unity gain or amplify the signal just enough to make up for line losses.

Coax cables come in two commonly used sizes, RG-59-U and RG-11-U, the commonly used designations. Both have impedances of 75 ohms, which is standard for video equipment. RG-59-U is the smaller cable, with higher losses but perfectly acceptable for runs up to several hundred feet. Various versions of this cable are available, some with low-loss insulation (foam) or double shielding. Normally, the regular version is acceptable for video. Two types of connectors are used, BNC, and UHF. The type selected usually depends on that required to mate with the equipment connectors. Adapters to convert from one type to the other are available. The BNC is a bayonet type and the UHF screws on the jack. The larger RG-11-U uses only the UHF type connectors. Another cable type has a solid aluminum shield and comes in several diameters. It requires special connectors.

The foregoing information applies to normal studio video wiring as well as video distribution lines. In the video system interconnections, as opposed to distribution on long lines, the smaller cable is usually used.

Balanced video cable requires special amplifiers to convert from balanced to coax cable and vice-versa. Special connectors are also required. The balanced line impedance is 124 ohms, incompatible with equipment designed for 75-ohm unbalanced cable. Thus, the need for special amplifiers. Equalizing amplifiers are also used, but these are the same as those previously mentioned, and are found in the 75-ohm side of the circuit. Multipurpose amplifiers are available to accept either a differential or the grounded shield input of coax lines or balanced lines, depending on the input jack used. Usually, equalizing circuitry is included. Other receiving and sending systems contain separate modules—balanced to unbalanced, unbalanced to balanced, equalizing amplifiers and a power supply—all usable in any combination with a plug-in frame. Figs. 9-1 and 9-2 shows one such system.

RF DISTRIBUTION SYSTEM

RF distribution can be very complicated in a 12-channel system, or relatively simple if only a few RF channels are used. A full 12-channel system needs critical mixing equipment, as well as bandpass filters and amplifiers. This is needed to connect all twelve VHF channels to one output line without interchannel interference. If you intend to use such a system, you should ask for professional help. Have the system installed on a bid basis, with very good specifications stipulated in the checkout procedure. Be sure you know what to specify, and do it in great detail. The information you use to determine specifications should be as up to date as possible because the state of the art in such systems is constantly being up-dated.

If you plan to originate only a few channels, you can probably do it yourself. Don't use adjacent channels, and you probably won't have any problems. If you want to originate only two or three channels, use Channels 2, 4, 6, 7, 9, 11, or 13. (6 and 7 are not adjacent to each other.) If you need more channels than that, you would be better off relying on professional help, unless you are very familiar with CATV head-ends and RF mixing networks.

Fig. 9-1. Shown is a balanced line driving terminal with pre-equalization. (Courtesy Dynair)

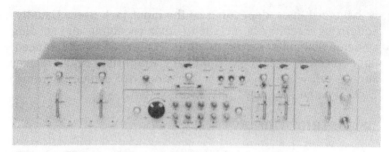

Fig. 9-2. This is a balanced line receiving terminal which includes sophisticated equalizing circuitry. (Courtesy Dynair)

In some cases you might want to distribute the local CATV signal in a school, with modulators to originate local programming on the vacant channels. If the local CATV feed has a sufficient number of vacant channels, this can be an ideal set up. Be sure the modulator you select has steep skirts, to avoid interfering with adjacent channels. If you run into such a problem, you can add a channel amplifier and-or filter of the correct bandpass to match the modulator. This will generally take care of any problems. The added RF signals should be the same level as those on the cable at that point. If just a few signals are to be added, a resistive splitter hooked up backwards will do the job. More about mixing and splitters later.

The modulator mentioned is used to convert video and audio to a VHF TV channel, acceptable to any TV set. A separate modulator is used for each channel. Some contain bolted down printed-circuit boards and cannot be converted to another channel. Others are modular in construction and are easily converted to any frequency by replacing the output module. In some models the module can be replaced from the front; with others the unit has to be removed from the rack. Some modulators have a meter on the front panel to set the audio and video modulation levels. This is a very useful feature and is highly recommended. Other modulators require external test equipment to adjust the modulation levels, and some have to be dis-assembled to make the adjustments. With consistent input levels the modulator adjustment needs to be made only during the initial set up when installed. However, it is nice to be able to easily check this adjustment if the signal doesn't look right. Aside from the front panel meter and modular construction, modulators are much alike. As mentioned before, it is desirable to have a modulator with steep skirts, so additional amplifiers and filters will not be required. Fig. 9-3 shows a typical modulator.

If only a few channels are originated, they can be mixed with resistive splitters, as stated earlier, or with directional couplers. In the latter case, the modulators feed into the taps and the directional couplers are connected together. The input connection on the first directional coupler is terminated and the others are connected output to input. A mixing network for adjacent channels is a complicated system, and more information than I can impart here is required for success. A few simple hints: Don't mount adjacent or harmonically related channels next to each other in the rack. Also, be careful of cable positioning in the mixer system because it is very critical. It has been said that solid jacketed aluminum sheath cable is best for this purpose. It will not move easily when placed in position and it has very good shielding characteristics. A poor mixer design will result in readily apparent interference between channels. Vertical and horizontal sync bars will float through the pictures and even a faint picture will be noted in the background of other channels. This is called windshield wiper interference because of the effect produced by the interfering horizontal blanking bar as it passes through the picture.

Another device you might find useful is a channel converter. This allows you to take a given VHF channel and retransmit it on any other channel.

SPLITTERS, DIRECTIONAL COUPLERS & OTHER PASSIVE DEVICES

Passive parts in the RF distribution system such as splitters, directional couplers, wall outlets, pressure taps,

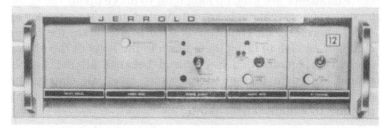

Fig. 9-3. This modulator uses easily replaceable plug-in modules. (Courtesy Jerrold)

attenuators and equalizers are commonly used. A perusal of the catalogs will reveal many other devices for special purposes.

A splitter is commonly used to divide a signal. Both 2- and 4-way splitters are available; that is, a single input to two or four outputs. The 2-way type has a 3 dbmv loss, the 4-way has a 7 dbmv loss. A splitter can be used to split a trunk line to two buildings, or a trunk line in a building to feed two floors, to give just two examples.

A directional coupler will split off a signal with only a 0.5 dbmv loss on the main line, while the split-off signal is down 8 dbmv or more, depending on the coupler chosen. The advantage of the directional coupler is that the split-off signal has very little effect on the main trunk line. Thus, the split-off signal can be opened or shorted with little or no effect on the trunk signal which is feeding other areas.

Wall outlets are used to feed the signals into rooms. You can get self-terminating types for use where no cable is connected to them, and some have built-in attenuators of various values to reduce the signal level. They are designed to mount in electrical receptacle wall boxes or can be surface mounted.

Pressure taps are something like directional couplers. They mechanically attach to the trunk line, piercing the jacket, shield and insulation to make contact with the center conductor. They have various values of built-in attenuation. A possible use might be to run a cable down a corridor ceiling to act as the trunk line. Pressure taps could then be used to tap off the signal to feed rooms along that corridor. Splitters or directional couplers also serve a similar purpose.

Attenuators come in various values; for example 3, 6, 10 and 20 db. They are used to reduce the signal level when required.

Equalizers are used on long lines to make up for the cable losses. The high channels are attenuated more than the lower ones. Therefore, an equalizer is used to attenuate the lower channels enough to bring them down to the level of the higher ones. This equalization is required because all the signals should be fed to the set at the same level. Of course, the equalizing network has to be placed at a proper distance along the line to exactly offset the losses.

Incidently, the ideal level at the TV set should be +4dbmv, give or take a few dbmv. This is the level to shoot for at each room outlet. Now you can see why attenuators, room outlet connectors with built-in attenuators and directional couplers with various attenuation values are available.

Other useful passive devices are available, too. For example, you can get a splitter that will divide the input into high- and low-band VHF signals; that is, Channels 2 through FM come out one tap and 7 through 13 come out the other.

You can buy passive filters which allow the passage of a single channel while attenuating all the others. This comes in an active version, too, called a channel amplifier. Other traps and filters are also available for special purposes.

RF AMPLIFIERS

Many types of amplifiers are used in RF distribution systems. Bridging amplifiers, building amplifiers, trunk amplifiers, and many others are available for different requirements. Some include directional couplers and equalizers. Others can selectively amplify high- and low-band signals, which gives some effect of equalization. Weatherproof cases are available for outside mounting. The amplifiers are sometimes powered locally with 115v AC or they can be powered by a supply located at a remote location. The power is fed over the same cable used to distribute the signal. A typical amplifier is seen in Fig. 9-4.

The spacing between amplifiers depends on the model. It is stated in dbmv. That is, after the signal has declined in amplitude a certain amount, let's say 16 dbmv, for example, another amplifier is required. Signal loss is caused by splitters, couplers and cable losses. The loss in passive devices is stated in the specifications and is referred to as insertion loss. Cable loss is also stated in its specification and varies with frequency. It is usually stated in dbmv per 100 ft.

RF CABLE

Various types of cable can be used for RF distribution. Foam (cellular polyethylene) dielectric is preferred; its loss is much lower than solid dielectric types. RG-59-U type cable is usually used for drops from the trunkline to the individual outlets. Truck lines can be RG-11-U or solid aluminum sheath cable. The solid aluminum sheath type requires special

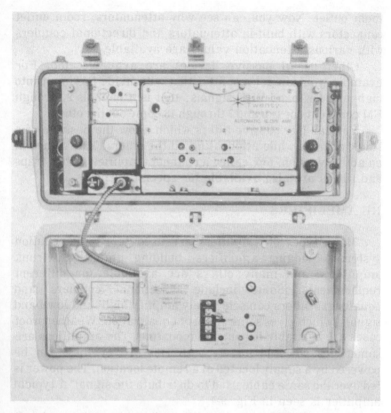

Fig. 9-4. This is a CATV amplifier with built-in RF equalization. It is operated by an external power supply. It also contains AGC circuitry to help keep the gain constant regardless of ambient temperature. It is mounted in a weatherproof case. (Courtesy Jerrold)

connectors. A special tool to mount the connectors is a useful accessory, though a hacksaw can also be used to prepare the cable. The connectors and special tools are usually available from the cable manufacturer. This type is recommended because it is weatherproof and has very low attenuation, thereby requiring fewer amplifiers on the trunk line. The fewer amplifiers you use, the less chance of failure and the lower the deterioration of the signal and noise.

TEST EQUIPMENT

Some special test equipment is useful to calibrate, adjust and check an RF distribution system. A field strength meter is used to check the level of each of the VHF channels. It can measure the audio and video carriers separately, as well as FM signals on the cable. It can also be used to orient TV antennas for maximum signal pickup. Fig. 9-5 shows a typical field strength meter.

When the system is initially set up and operating to your satisfaction, log the levels of each channel at each amplifier output and room outlet. Any other suitable points you think critical should also be checked, especially the outputs of any tunable filters or traps used in the system. Also log the output level of each modulator and channel converter. This is a time consuming endeavor, but the information accumulated will be extremely useful when you have level problems in the system some day. Test taps can be permanently connected to the points you wish to measure. In physical appearance it resembles a directional coupler. With it you make test readings without disconnecting any RF lines.

Another piece of test equipment which is desirable is an RF sweep generator covering the VHF channels. It should be reasonably flat in frequency response over the frequencies of interest. Any variation from a flat response should be noted so

Fig. 9-5. A good example of a VHF field strength meter. (Courtesy Jerrold)

allowances can be when making measurements with it. A marker generator should be used with the sweep generator to mark band edges or any other point of interest. If the markers are added after the sweep is generated, the marker "birdies" will be sharper and more easily recognized on a scope. Obviously, the sweep generator is used to be sure there are no peaks or valleys in the frequency response of the system.

One final piece of equipment is a time domain reflectometer (TDR). It is used to check for any variance in the impedance of the cable system. Its sensitivity will show every connector defect or discontinuity in the system. Kinks in the cable, smashed sections, loose connectors, or any other fault will be readily apparent. It is especially useful for checking cables after they have been pulled through conduit to be sure they haven't been damaged. Incidently, a TDR is used to check only passive devices and cable; it is not used for amplifiers. It will also show any manufacturing faults in the cable. It is nice to have around all the time, but is especially useful to check a newly installed system. If you can't afford to buy a TDR, you might rent one long enough to test the system right after installation. It uses a chart recorder or scope to graphically analyze the system. The TDR is accurately calibrated and will pin down exactly how many feet down the cable the fault is located. Incidentally, all unused drops and taps in the system should be terminated to avoid ghosts or reflections in the picture.

Chapter 10

Audio Mixing Consoles

An audio console is an almost indispensible part of any system using more than one or two microphones. Essentially, an audio console will accommodate a large number of audio inputs and blend or mix those selected into a single output. The simpler audio console has only a few inputs and a single output, while the type used for recording master tapes when making records can have 24 or more inputs and 16 or even 32 output channels. For television, a console with 18 inputs and two outputs is considered a large one.

A little audio terminology might be in order before we go too much further. Inputs are either high or low level. High-level input impedances can be 600 ohms or 20,000 ohms (nominal) and the levels can be -20 dbm or 0 dbm. A low-level input is usually 250 ohms (nominal) and will accept a -55 dbm level signal. The low-level input is normally used for low-impedance microphones and the high-level inputs for most other sources. A mixer is the level control for the input signal. A VU meter is the level indicator calibrated in percent modulation and dbm. Usually, zero dbm is equal to 100 percent modulation on the meter. Normally, the mixers are set to keep the level at a maximum reading of 0 dbm on the meter. The actual output level of the console at an indicated 0 dbm can be 0, +4 or +8 dbm, depending on the console design. A separate VU meter should be provided for each console output channel.

The mixing console is normally used in conjunction with the production switcher; the switcher selects the video source and the audio console provides the audio to go with the video.

TYPICAL CONSOLES

Two basic types of consoles are in general use. One type has a front panel with fixed placement of the level controls, mixer keys, VU meters, and other hardware. Inside this console, high- and low-level preamplifiers may be fixed or

Fig. 10-1. A typical small "off-the-shelf" audio console with eight mixer inputs. (Courtesy QRK Electronics Products)

hard-wired into the console, thereby predetermining the number and types of inputs that can be used. In other versions, the mixing channels can be either high- or low-level types, selected by plugging in either a preamplifier or a high-level input module. This system allows some flexibility in the capabilities of the console. The low-level input switches, in either case, are usually wired to mute or turn off the speakers in the studio to avoid feedback when the studio microphone is turned on. A typical small console is shown in Fig. 10-1.

The other type of console is considered a custom model where the operating panel is set up physically to the purchaser's specifications. Any number and combination of high- and low-level mixers will be provided, as well as any number of output channels and VU meters. The mixer channels in this type of installation can be of many types. For example, the simplest has only a gain control and possibly an attenuator switch which allows the mixer to be used for either high- or low-level inputs. An echo or reverberation level can also be provided, with a switch to select pre or post echo. In addition to the previously mentioned controls, an equalization network can be selected for that channel, controlled by potentiometers (mixer gain controls) in the plug-in module. Any combination of the above is possible, plus many other options limited only by the imagination—and bankroll. A very large console of this type can be seen in Fig. 10-2.

In addition to the above controls, a custom console can also include monitor attenuators. These control the monitor amplifiers and speakers, which can be switched to monitor any of the console outputs, or a separate monitor and speaker can be provided for each channel. A separate monitor or monitors should also be available to feed a studio speaker or speakers.

Normally, a television console has only one output in use at any one time. The second output channel can be used for a spare, an audition bus, or re-entered into one of the mixers and used for a submaster. The submaster allows you to preset a group of mixers, feed them into the unused output bus, and feed that back into the bus on the air through another mixer. This allows fading the entire group fed to the submaster in and out with a single control.

Another use of the second bus is as follows: If the studio muting circuitry can be provided with a switch so it will mute only if the first output bus is fed to the studio, any accompanying sound effects or music can be fed to both output buses, with the studio mikes being fed to the first bus only. This permits the second output bus to be fed to the studio monitor, allowing the talent to hear all the sound effects,

Fig. 10-2. A large 2-channel custom audio console. (Courtesy Audio Designs)

music, or possibly voice to enable them to "lip sync" without any risk of feedback. Since the microphones are not set to feed to audio output channel 2, feedback is not possible. The first channel, which includes all the audio sources, is then fed to the transmitter or VTR. The console must be able to feed the mixers to either or both output channels to use this feature. This is usually called foldback.

A cue amplifier should also be provided. This is used to cue records and tapes to the desired starting point. The cue amplifier can be used by setting an attenuator to its maximum position or a switch on the mixer module can switch that mixer output off the output bus and into the cue amplifier. It is also possible to provide a switch on the console to use the cue amplifier for a monitor amplifier on either output bus. This gives the console a little more redundancy. Obviously, as you've probably noticed already, I am a great believer in redundancy. It can eliminate a lot of unnecessary troubleshooting and provide immediate on-the-air use of secondary equipment. The cue amplifier attenuator should also be a part of the operating panel. In some inexpensive consoles an outboard cue amplifier may have to be furnished. Each output channel should also have an attenuator, as well as a VU meter.

The off-the-shelf, as well as the custom console, can be equipped with a multiple input, single output switch to provide a large choice of inputs to one or two mixers. This will reduce the number of mixers required for a large number of inputs with a slight sacrifice of flexibility. Several types of switches can be used; rotary or pushbutton switches can be used directly, or remote controlled reed relays can be operated by either type of mechanical switch. The reed relay system is less likely to be plagued with dirty contacts and the resulting poor audio.

The custom console can also be purchased with a built-in audio patch panel system (more later) and, if desired, remote controls for the audio tape equipment used in the system.

We find it best to keep all high-level inputs at the same impedance and level. This allows audio to be patched around without worrying about impedances and levels. If all audio inputs in the system are balanced bridging types and all audio sources are directly terminated at their outputs, it will be very difficult to inadvertently upset audio levels. You can bridge up to three audio inputs across each line, and if the input impedances are significantly greater than 20,000 ohms, one or two more. A rule of thumb is that the total impedance bridged across a 600-ohm line should be at least 10 times (6000 ohms) the line impedance. If all the input impedances are equal, the total impedance can be easily calculated by dividing the bridging impedance by the number of inputs. If the line is loaded excessively (too many inputs) the output level will be lowered and the frequency response will also suffer. Also, a VU meter across that line will no longer read correctly, since it is calibrated to indicate levels on a 600-ohm line. We like to use 0 dbm as a nominal audio line level, with the exception of the microphone levels of course. All audio equipment can be purchased to provide 0 dbm. Try to avoid using -20 dbm in the system; though there is nothing inherently wrong with such equipment, we feel it is simpler to keep all the levels alike—at 0 dbm.

Unfortunately, most off-the-shelf audio consoles have 600-ohm inputs. In that case you could feed the audio console with one output of an audio distribution amplifier, terminating the other outputs at the source and bridging across them instead. Another possibility: If the console has 0 dbm 600-ohm inputs, you can use the console as the termination and bridge across that line. Of course, this is not so convenient as using all bridging inputs. If the console uses -20 dbm inputs, a 20,000-ohm to 600-ohm bridging transformer can be used on the input, thus converting it to a bridging input.

Another point to keep in mind is "head room." This is the maximum instantaneous level that can be provided by a piece of audio equipment. If the maximum output available is +10 dbm to +20 dbm, distortion is much less likely to occur on an occasional high-level peak in the audio. The console VU meter does not read audio peaks; it is damped so that it reads the mean level based on an average over a period of time. The amplifier should be capable of reproducing those unseen peaks without distortion. Even if the peak is later eliminated by a limiter, the other audio frequencies being reproduced at that time will not be distorted, as they would be if the audio peak was clipped in the console amplifier.

Consoles are built-in table-top designs or in floor model frames. The amplifiers and preamplifiers can be built into the console, or rack mounted elsewhere with the console acting as a remote control center.

A somewhat different type of audio console is available. This is much easier to operate than the normal console. It is designed for use in fast radio formats but can also be used for television. If inexperienced operators are to run it, it would be worth looking into. It doesn't use any external mixer level controls, just cue, line and monitor gain controls. Each mixer channel has a switch to select either of two inputs and four pushbuttons. The buttons are labeled start, stop, override and cue.

The start button puts that channel on the air; the stop button takes it off the air, and the cue button puts it on the cue bus. The override puts it on the air at normal level while reducing the other channel(s) on the air by a set amount. This is used to crossfade or for voice-over production. When a start button is pressed, all the other channels are automatically switched to stop.

The start and stop pushbuttons have external contacts which can remote control audio sources such as ATRs, CTRs, VTRs, turntables, etc. Thus, when the start button controlling the turntable input audio is actuated, the turntable can also be made to start as well.

Internal controls are used to set the gain of each mix channel. Other internal controls are used to set the amount the other channels are reduced when the override switch is used. One output channel is available. A photograph of the above described console can be seen in Fig. 10-3.

MONITOR AMPLIFIERS

Audio monitor amplifiers used should be of execellent quality and placed at strategic points in the system where the audio quality can be monitored. Typical locations may be at the audio console and at the program or distribution switcher.

A good audio monitor should have several characteristics: A reasonably flat frequency response curve is essential. The amplifier should be flat within 1 db, preferably better. Usually, the response is measured over a range of 20 to 20,000 Hz. With the present state of the art in transistorized amplifier design, this is an easy specification to meet. The amplifier should be a high-powered unit, capable of driving the popular low-efficiency speaker systems and to keep the output in a power range that will hold the inherent noise at a low level.

This also leaves plenty of power available for large excursions such as those generated by sharp transients in the programming. This makes the resulting output sound much more open, crisp and clean. The amplifier should be easy to service, with the components easy to reach and replace if necessary. The transistors used should run well within their ratings to preclude frequent replacement. The output circuitry should be protected to avoid burning out the power output and driver transistors if the speaker terminals are inadvertently shorted or opened when the amplifier is running at full power. A number of power amplifiers of various types are shown in Fig. 10-4.

Tone controls are not required; if present they should be run in the flat position, otherwise a loss or increase of high or low frequency response would not be noticed. The purpose of a flat power amplifier is to make any such deviations noticeable, thus tone controls defeat the flat frequency response you paid for. It might be argued that the tone controls are available to make up for deficiencies in the reproductive capability of the room where the speaker is located. It is best to try to tailor the room acoustics for a flat frequency response or play enough material through the system so that you are familiar enough with the sound at the

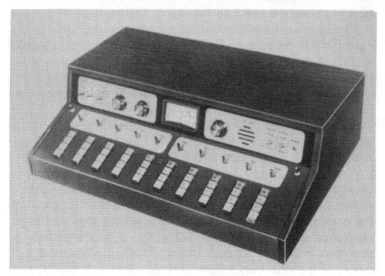

Fig. 10-3. This is the CCA simple-to-operate console. Notice the absence of external mix level controls. (Courtesy CCA)

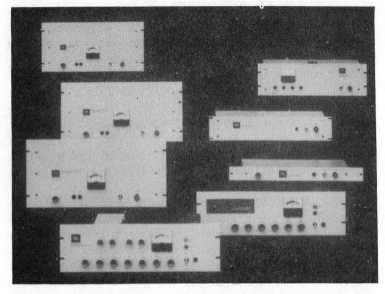

Fig. 10-4. A representation of power and PA amplifiers available for almost any purpose. (Courtesy JBL)

correct frequency response to recognize anything substandard. This is better than adjusting the tone controls whenever things don't sound just right. If you do that, you can never be sure of the true sound of your system.

Speaker systems come in numerous sizes and prices. An example can be seen in Fig. 10-5. Read the manufacturer's specifications and pick the one with the best characteristics in the price range you can afford. Some of the speakers in which you might be interested are available in the larger high fidelity shops. You could listen to them there to find the one with the sound you prefer. This is an artificial environment in which to test speakers, but it is better than picking one by just reading the specification sheets.

Speaker systems come in many sizes, shapes and types. Incidently, when we refer to speaker systems, we mean a speaker or speakers, usually the latter, mounted in a matching enclosure. It is best to buy the speakers mounted in an enclosure. If you buy them separately, they may not be designed to work with each other, although when used with other components they might work admirably. Up to a point, the larger the enclosure, the better the low-frequency response. Shapes vary, and are usually unimportant if the radiation pattern is suitable for the room and speaker placement you have in mind.

There are several basic types of systems—acoustic suspension, bass reflex, folded horn, and many others. If the speaker has tone controls, these, as in the amplifier, should be set to the flat position.

Fig. 10-5. A pair of professional quality speaker systems. The speakers used in these systems are also shown. (Courtesy JBL)

Speaker system impedances vary, and should be selected to match the amplifier output, or vice-versa. Most amplifiers will match any of the common speaker impedances—4, 8, or 16 ohms. Some amplifiers will only develop full power into the proper speaker impedance, although they will operate at somewhat reduced power into other loads. This should be taken into consideration when you choose an amplifier and speaker system.

Some speakers require a 25- or 70-volt feed from the amplifier. This system is usually not used for critical applications; it is more suitable for public address type systems such as paging or covering a large area with a number of speakers. The advantage of such a feed is that a number of speakers can be connected in parallel and each speaker can be adjusted to a specific loudness level by selecting a tap on the matching transformer. Thus, the amplifier output is not impedance sensitive, but will accept any number of speakers connected to its 25- or 70-volt output up to the maximum power available. Connecting or disconnecting speakers should not change the level of the others connected to the amplifier.

Other areas where audio monitoring is desired might be better served by smaller less expensive amplifiers and speakers. Such areas could be where the audio content is more important than quality. If built-in monitor amplifiers are not provided, they can be added without too much trouble. Small, low-powered transistor amplifiers and speakers are widely available for such use. They can be built into the equipment or mounted on the back of blank rack panels to be used where desired and bridged across the 600-ohm line output. Be sure to provide bridging transformers for unbalanced inputs. You would be amazed at how much crosstalk, feedback, and other problems can be caused by a few unbalanced inputs in an otherwise balanced audio system. Of course, if all the inputs and outputs are unbalanced, this is no problem.

The speakers used for such systems can be small replacement type speakers mounted behind a blank panel or at some other convenient point. If mounted behind a blank panel, a number of holes can be drilled in the panel to let the sound out. Enclosures are not really necessary as we are interested in content, not quality. These little amplifiers put out only a half watt to one or two watts, so almost any speaker will be suitable. Don't worry too much about using the exact speaker impedance. For at such low levels, it probably won't matter enough to worry about.

Such monitors are used for VTRs, film projectors, and any other point where you wish to monitor audio content. One- and 2-inch VTRs usually have built-in monitors while half-inch VTRs usually do not. Audio monitoring is usually useful at the VTR for cueing and verifying that the audio circuitry is working properly, especially while recording.

The film projector sometimes is equipped with an audio monitor, although it is not essential because the film is usually cued with numbers on the leader. It is nice, however, to be sure the exciter lamp and audio circuitry is functioning properly before you have to roll the film.

The studio, of course, requires a monitor speaker. It should be of good quality if the studio is used to screen programs after they have been recorded. The amplifier driving this speaker can be supplied by a distribution switcher (preferred) or the audio console. If a distribution switcher is used, the video monitor should be supplied through the same switcher. The studio speaker must be muted when the microphones are turned on unless a foldback circuit is used, as mentioned earlier.

Utility monitors can also be located at points where it might be necessary to listen to the programming, such as the telecine room. These monitors can be bridged across the audio console to monitor a production or across the program switcher to monitor the program on the air. It might be useful to have a switch to select either of these two sources, or both could be monitored using separate monitors and speakers, with the one you are not interested in at the moment turned down.

Chapter 11

Audio Program Sources

Audio recorders come in a wide variety of operating formats—reel-to-reel, cartridge and cassette types. Playback equipment includes the above tape machines and disc playback units. Individual studios rarely use disc recording equipment now, so this type of equipment is not covered.

Reel-to-reel audio tape recorders, we'll call them ATRs from now on, are the old standby for studio operation. They are normally used for recording long pieces of background audio, audio copies of television programs, musical numbers to be lip-synced by a performer, and other programs too long to be conveniently recorded on a tape cartridge. The ATR comes in five formats; full-track monaural, half-track monaural, half-track stereo, 4-track stereo and 4-track quadrasonic. For television use, usually a full- or half-track monaural machine is used, full-track preferred. We have been discussing the ¼-inch tape machine; ATRs are available for ½-inch, 1-inch, and 2-inch tape, too (with many more than four channels), but are seldom used for television.

ATRs are available in many price ranges, with two or three heads, both local and remote control capability, and in amateur or professional designs. A professional ATR is shown in Fig. 11-1.

The typical 2-head machine has record-playback and erase heads. One preamp is used in both the record mode and the playback mode. This means that you cannot listen to a program as it is being recorded. With the 3-head machine (erase, record and playback) two preamps are used, one for recording, the other for playback. Both can be used at the same time, allowing the signal recorded on the tape to be monitored as it is being recorded.

Local control ATRs use mechanical linkages to switch between stop, play-record, fast forward and rewind modes. A recorder designed for remote control uses solenoids to change

Fig. 11-1. This is a typical professional monaural ATR. (Courtesy Telex)

modes. The solenoids are operated by pushbuttons located at a remote point.

Now comes the question, what is the difference between an amateur and a professional ATR? I could say price and performance, but when you read the specification sheets, the amateur machines seem to equal or exceed the published specifications of the professional machines. Probably the distinguishing mark of the professional machine is reliability, repeatability of published specifications after prolonged use, and mechanical rigidity. You have to pay a large premium for a professional machine, but if that is what you need for your requirements, it should be your choice. If you use an ATR only a few hours a week, you could probably get by with an amateur quality ATR. Incidentally, the published specifications for a professional machine are usually very conservative, while amateur machine specifications usually are predicated on top performance.

Dolby circuitry is available for both professional and amateur ATRs. This feature, if you are not familiar with it, reduces the inherent noise generated by the recording process which especially degrades the high frequencies. It is a useful addition to an ATR. However, tapes recorded on a machine using the Dolby process should also be played back on a Dolby equipped machine to retain the proper frequency response.

The ATR selected should have a balanced high-level bridging input. Some also have a low-impedance microphone input, which can be used for remote location recording. The most popular speeds are 7.5 or 15 IPS (inches per second). The 7.5 or even 3.75 IPS speeds are quite acceptable for speech, but 15 is prefered.

Equalization controls for recording and playback should be provided, as well as record current adjustments to optimize the bias setting for various types of tape. A single brand and type of tape should be selected and used as a standard. The ATR circuitry should then be adjusted for optimum performance with that tape (see the tape recorder manual for the procedure). Remember, there is no sense in buying an expensive recorder if you are not going to allow it to perform as well as it can.

CARTRIDGE TAPE MACHINES

A cartridge tape recorder (CTR), Fig. 11-2, is a useful piece of equipment for short pieces of recorded material, such as station identifications (IDs), program themes, sound effects, commercial and other announcements. This machine uses a cartridge of tape available in different playing lengths. CTRs are usually half-track format machines, with the program material recorded on one track and cue pulses recorded on the other. Stereo machines are also available, but are not usually used for television. The cue portion of the tape is used to record a pulse at the beginning of the program material to cue the tape. In operation, the CTR records a stop pulse at the beginning of the recorded material. When the tape is played back, it plays the recorded material until it detects the stop pulse. It then automatically stops. The cartridge can then be removed to be inserted and played again later. Pulses other than the stop pulse can be recorded on some machines to energize relays, the contacts of which can be used to operate slide projectors, film projectors or other equipment.

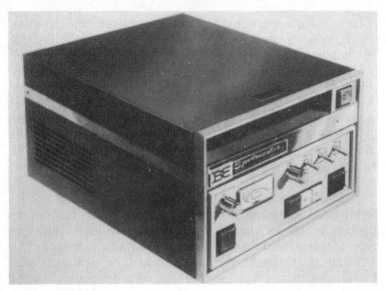

Fig. 11-2. This is a popular cartridge tape recorder—playback unit. (Courtesy Broadcast Electronics)

The CTR record section is usually purchased as a separate unit and can be connected to any playback unit, converting it to a record-playback unit. Sometimes playback units are available in pairs on a single chassis. When a dual playback unit is connected to a record amplifier, either playback unit can be used to make a recording, or even dub from one playback unit to the other without external patching. A CTR does not usually include an erase head, so tape cartridges must be erased with a bulk eraser before they are used for recording.

CASSETTE MACHINES

Professional quality cassette machines are becoming available now. They are similar to those so widely used for entertainment, but are more reliable and have much better electronic specifications. The professional models usually run at a higher tape speed, 3.75 IPS as opposed to 1.825 IPS, resulting in better high-frequency response and less inherent noise.

For the operator on a limited budget, amateur cassette recorders can be used with external transformers to provide compatibility with the impedances offered by other audio equipment. This means converting the input impedance from high-impedance (hi Z) unbalanced to hi-Z balanced and the output from hi-Z unbalanced or speaker (4 ohms balanced) to

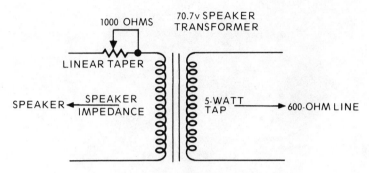

Fig. 11-3. This circuit can be used to convert a speaker output signal to a 600-ohm line feed.

600 ohms balanced. A method of converting any speaker output to 600 ohms is shown in Fig. 11-3. Compatible signal levels also have to be determined, of course.

DISC PLAYBACK EQUIPMENT

Disc playback equipment is probably used the most. Again, you can choose between amateur and professional turntables and tone arms. The professional equipment provides the most reliability and, in this case, has other more important advantages. A professional turntable will reach operating speed in about a quarter turn of the platter, allowing a phonograph record to be easily integrated into a program with no pause while the record comes up to speed. A professional turntable is easier to back cue, which involves rotating the turntable backwards by hand, past the start of the desired recorded material by a sufficient distance to allow the platter to reach the proper speed before the recorded material begins. The stylus must be at the correct angle, determined by the cartridge and tone arm adjustments, to allow the record to be rotated backwards without damaging it.

A rim-driven turntable is usually the best for a fast start. The starting process must be smooth enough so the stylus,

Fig. 11-4. A typical professional 2-speed turntable and tone arm. (Courtesy Gates)

with a nominal stylus pressure setting, does not jump to an adjacent groove. Slip starts eliminate undesired pauses or slow starts. To execute a slip start the record is held while the turntable comes up to speed, then the record is released. This requires a mat between the record and platter that will slip freely on the platter, but will adhere well enough after release to preclude slippage of the record while playing. Felt works very well. If the turntable will reach the correct speed quickly, as previously discussed, slip cueing is not necessary.

Turntable manufacturers usually recommend a tone arm that will work well with each turntable they offer. A complete system is shown in Fig. 11-4. The pickup cartridge should be a durable unit that allows easy stylus replacement. A stereo cartridge should be used so any record can be played without damaging the record. A diamond stylus is best to use and it should be checked frequently for wear. The outputs of the cartridge can be paralleled so a monaural preamplifier can be used, or a stereo preamplifier can be used with its outputs paralleled to provide a monaural output.

A monaural preamp is considerably cheaper, but if you have a stereo cartridge you have the option of using it for stereo reproduction later if you choose a stereo preamp. Also with a stereo preamp you have two monaural preamps if you parallel the cartridge outputs. This gives you some redundancy if it is ever needed. To keep all the levels compatible, the preamp output should be 0 dbm at 600 ohms.

Equalizer circuitry is usually furnished with the turntable preamp to cut the high frequencies if the record is scratched.

The turntable is usually mounted in a pedestal. Sometimes CTRs and other equipment are also mounted in the pedestal. The pedestal should be firmly mounted to the floor to reduce possible vibrations. It can be filled with sand to help stabilize it if necessary.

MICROPHONES

When you start thinking about microphones you have a lot of choices—hand mikes, stand mikes, table mikes, boom mikes, lavaliers, cardioid, super cardioid, bidirectional, omnidirectional, dynamic, ribbon and condenser types. Perhaps it might be best to describe the acoustical and electronic properties of the various types of microphones so you can be in a better position to decide which type would be the best for your requirements.

The omnidirectional mike has equal pickup sensitivity in any point around the microphone. The cardioid sensitivity is the greatest at the front of the microphone in a roughly heart-shaped area (hence the name), with the point of the pattern oriented towards the microphone. The super cardioid is a modified cardioid pattern, with less sensitivity to the sides and more on the microphone axis. The bidirectional mike has a figure eight pickup pattern, with the center of the pattern at the microphone. The pattern extends to the front and rear of the microphone, positioned vertically. Each lobe is similar to a cardioid pattern.

The dynamic microphone is operated by sound pressure on the pickup element, together with tuned ports and other acoustical devices in some types to adjust the frequency response and directionality.

The ribbon mike is somewhat fragile compared with the dynamic; it can be damaged by wind or severe sharp sound transients such as gunshots. A thin ribbon is moved by sound pressure between two magnetic poles to generate the signal.

The condenser mike is excited by the change in capacitance between the pickup elements when one element is moved with relation to the other by the variable air pressure of the sound. The condenser mike is considered the best sound transducer, but it requires a separate power supply and is usually quite expensive.

The ribbon mike is used for studio applications. Some can be adjusted for any of the pickup patterns mentioned. This is called a polydirectional microphone. The RCA 77DX is a good example (Fig. 11-5).

You are the best judge of the microphones that are best suited for your purpose. We would recommend a few small lavalier mikes (small microphones designed to be suspended around the neck on a cord), and several mikes that can be mounted on a floor or table stand or be hand held. The Electrovoice 635A is a good example. If a boom mike is desired, a super cardioid is a good choice. A 635A is shown in Fig. 11-6 and a boom mike in Fig. 11-7.

If there is a lot of ambient noise in your studio and your talent is more or less fixed in position, a highly directional microphone or the lavalier is a good choice.

The condenser mike is said to give the smoothest frequency response, with the ribbons and dynamics a close second.

The requirements and the acoustics of your studio should govern your choice of microphones. The final destination and treatment of the audio signal should also be considered. For example, it makes little sense to buy a very expensive condenser microphone with an essentially flat frequency response from 20 to 20,000 Hz and then record the resulting audio on a helical VTR with an audio frequency response of 75 to 10,000 Hz and a three or four db variance in the response curve. This is not to say the specifications of the input should not exceed that of the recording device or the ultimate reproducer of the signal; we only feel that a premium in price should not be paid for specifications that will be lost in the recording and reproduction of the signal. If those increased specifications can be purchased for a small increase in price, such as a better phono cartridge or better recording tape, the increase is

Fig. 11-5. The RCA 77DX polydirectional ribbon mike. (Courtesy RCA)

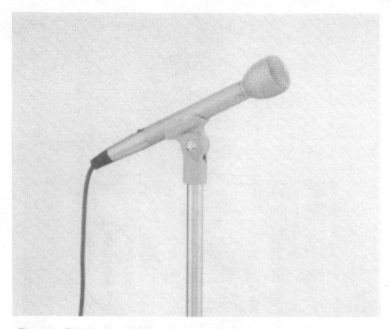

Fig. 11-6. This is an omnidirectional microphone which can be used on a stand or as a hand mike.

reasonable and should be accepted. After all, the better signal you have to start with, the better signal you will end up with. Consider the well known computer terminology, GI-GO (garbage in, garbage out).

In any case, you should use low-impedance microphones, which are much less likely to be affected by extraneous electrical fields, such as the 60-Hz power line. Even if your VTR or other audio input device requires a high-impedance input, a low-impedance mike should be used, with an external hi-Z to low-Z transformer at the end of the mike line. This will allow you to use a long mike line without sacrificing high-frequency response or picking up undesirable noise.

Another type of microphone is called the wireless mike. One type is a self-contained unit that looks like a large hand mike. Other types have a separate transmitter and antenna, worn by the talent under his or her clothes. A subminiature mike is used with this type. A wireless mike transmits a radio signal which is picked up by a nearby receiver and converted back to audio which is fed to the audio console. This type of mike allows the talent to move about the set without trailing a mike cord behind them. It has been received with mixed emotions by talent and engineers alike. If you need its unique capabilities, it is best to borrow a demonstration unit from your local dealer and see how it works in your situation.

MICROPHONE STANDS & BOOMS

Table stands can be purchased that will accept hand mikes, or special table mikes with built-in stands can be used. For example, with two people at opposite sides of a table, a bidirectional mike could be used. This must be positioned so

Fig. 11-7. A very directional microphone suited for use on a boom. (Courtesy Electro-Voice)

Fig. 11-8. A large boom with room for an operator on the platform. It is shown in the retracted position ready for storage. (Courtesy RCA)

that each person's voice results in the same microphone output level, otherwise the mike gain control on the console will have to be adjusted between the two speakers. If enough mike inputs in the audio console are available, a separate microphone for each person is best.

A floor stand mike can be fixed in the stand or held in a clip so that it can be removed by the talent and used as a hand mike for parts of the performance. We feel the hand-type mike is the most flexible. It can be used in a floor stand or table stand if required. This means fewer microphones will be required, since the one type can be used for many types of productions.

Mike booms range from small units, to very long booms mounted on a wheeled base designed to hold a mike over a piano to allow the pianist to speak or sing into it, see Fig. 11-8. The latter can be extended or retracted, raised or lowered, and the mike rotated on its axis during the course of a production. The wheeled base is large enough to accommodate a seated operator, but it is usually not moved about in the course of a production. A smaller boom is available, on a smaller wheeled stand, that can be moved about. It is counterweighted so the mike can be raised or lowered; it also allows the mike to be rotated for the best pickup angle. The boom can be extended or retracted, but this cannot be easily done while it is being used, as with the larger boom described earlier.

The shotgun or bazooka mike is used to pick up voices at a distance. It is highly directional and can be used, for example, to pick up questions from an audience. A less expensive version uses a parabolic reflector and a regular microphone. If you need a system of this type, try out the different kinds and see which works best for you.

Chapter 12

Peripheral Audio & Other Equipment

From the standpoint of the purpose served, audio distribution amplifiers can be compared with video distribution amplifiers. However, they will not be found as often because it is easier to bridge across an audio trunk line than a video line. If all audio inputs are bridging types and not more than three or four 20,000-ohm inputs are connected across the line, you will not need an audio DAs. But it is very useful if there are quite a few loads on a line. The audio DA usually is a unity gain device, although some have gain controls for a limited change in the output level.

PATCH PANELS

Audio patch panels are quite similar to video patch panels. The jack layouts, normals, grounding and other details covered in regard to video jackfields hold true for audio, too. An audio jack panel is seen in Fig. 12-1.

Two types of jacks are used—the common stereo ¼-inch phone jack and the double patch jack. The stereo patch (sometimes called CBS type) uses the plug tip and ring for the two balanced conductors and the sleeve for the shield. Phasing between stereo jacks must be carefully checked when wiring them because there is no way to change the phase later without rewiring. This can be an advantage or disadvantage, depending on how you look at it.

The double jack consists of two unbalanced monaural ¼-inch phone jacks mounted side by side. The patch cord plug has two ¼-inch plugs in one handle. Each tip connects to one of the balanced conductors and the two sleeves are connected together to the cable shield. The dual plug is notched on one side so the plugs can be phased alike. Thus, you can see how easy it is to reverse the phase on an audio line if desired. It is a good idea to get into the habit of always using the same hand to

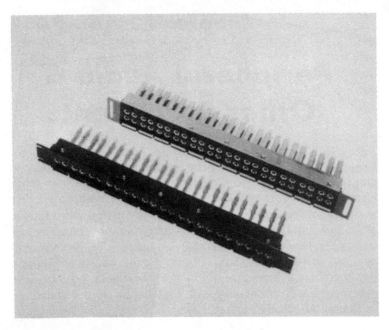

Fig. 12-1. Two audio patch panel types. The double-row type is more commonly used because it permits circuits to be normalled through in a natural manner. (Courtesy ADC)

insert plugs, with your thumb on the notches in the plug. That way, you will always maintain the same phase. See Fig. 12-2.

When wiring an audio patch panel, you can wire four jacks or pairs of jacks in parallel. This is a simple way to provide bridging connections across a line. In use an audio line is interrupted with two patch cords, which are plugged into the paralleled jacks (sometimes called a multiple). Then the other two pairs can be patched into bridging inputs. If a VU meter patch input is incorporated in the system, it can be connected to two pairs of paralleled jacks. As with the multiple jacks, any audio line can be interrupted and patched into the VU meter jacks. The jack panel is made so that the jacks are arranged in pairs, with too much space between adjacent pairs to put a plug into the wrong holes. An insulated panel should be used, so grounds can be separated if desired.

Traditionally, the patch panel jacks are wired to terminal blocks, called a "Christmas tree" or "pine tree" type because of the appearance from the side (Fig. 12-3). The blocks are mounted horizontally, usually, and the bottom surface terminals are wired to the patch panel jacks. The top surface terminals are then available for connection to the incoming and outgoing lines. Source terminations are usually connected to the terminal block.

The block should be wired in a logical manner. The terminals should be wired in the same order as they are positioned on the patch panel rows. Above all, never wire them haphazardly, or you will have a terrible time finding a given pair later. The phase should be kept alike on each set of terminals, too. The terminals are quite close together; therefore, guard against shorts caused by stray strands of wire or clipped off ends.

PROCESSORS

Several types of audio processors are used, including automatic gain control (AGC) amplifiers, limiting amplifiers and other special amplifiers.

An AGC amplifier will, to a reasonable extent, ride gain on the audio signal. The better ones will not increase the amplification level with no signal present, thus avoiding the amplification of only line noise. There are many types of AGC amplifiers, called by as many names, with differing characteristics. You will have to research manufacturer's specification sheets to discover which type will best suit your needs.

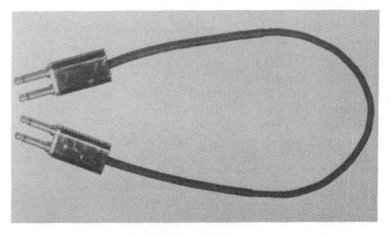

Fig. 12-2. A dual-plug type audio patch cord. You can see the notches mentioned in the text on one of the plugs.

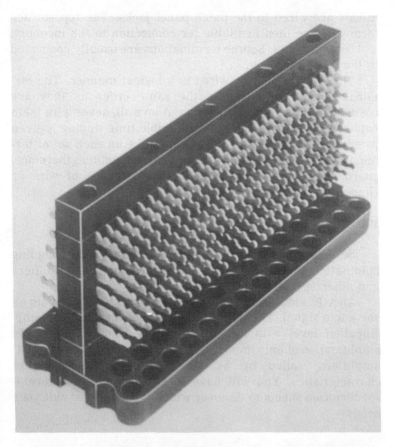

Fig. 12-3. A "Christmas tree" terminal block. (Courtesy ADC)

A limiter amplifier tends more to hold the signal to a maximum signal level. Some react to the peak of the signal waveform, others react to the RMS content of the waveform. Sometimes the AGC and limiter circuitry are combined in a single unit.

Special amplifiers are also available, such as a unit manufactured by Allison Research that offers all sorts of interesting effects. For example, one audio signal can be keyed on and off with the keying control provided by another audio source. This unit can also be set to eliminate any audio signal below a preset level; thus, it will wipe out any undesirable studio background noise such as air conditioner noise. This amplifier is shown in Fig. 12-4.

Various processing amplifiers are built by a number of manufacturers—CBS Laboratories, Allison Research, RCA and Gates Radio to name just a few. You can find many others advertising in such trade journals as:

> Broadcast Engineering
> 1014 Wyandotte St.
> Kansas City, Mo. 64105
>
> DB
> 980 Old Country Road
> Plainview, N. Y. 11803
>
> BM-E
> 200 Madison Ave.
> New York, N.Y. 10016
>
> Broadcasting
> 1735 De Sales St. NW
> Washington, DC 20036
>
> Audio Engineering Society Journal
> 60 E. 42nd St.
> New York, N.Y. 10017

There are others also dealing with these subjects. We think the above are among the most popular written with the professional broadcaster in mind. Some are technical while others deal more with the business end of television and radio. If you are not familiar with them, ask for a sample copy of each from the publishers and decide which suit your purpose. Some are free to qualified subscribers.

A test tone generator wired in an audio system serves a number of purposes. Two popular frequencies are 400 and 1000 Hz. A simple, continually running, fixed-frequency oscillator should be used. The output level should be 0 dbm. The oscillator can be hardwired into the audio console, and can follow the video test generator input into the program or distribution switcher. The tone generator output can also appear at patch panel jacks, permitting it to be patched into any desired input. It can be used to check levels anywhere in the system. It is also a good idea to record the tone on the lead

Fig. 12-4. The Allison Research "Kepex" audio processor amplifier eliminates all signals below a set level. (Courtesy Allison Research)

of video and audio tapes (not cartridges!), where it can be used to preset the level when the tape is played back. Some oscillators are available with several switch selected frequencies.

Reverberation units are valuable for certain purposes, as discussed briefly in the chapter concerned with audio consoles. There are three types of reverberation units available—those using a digital electronic delay, large sheets of metal or springs. The spring type is the cheapest by far. Whatever type is used, be sure its input and output impedances match your system.

A tape recorder, if it has separate record and playback heads, can be used for reverb or echo. This is accomplished by simultaneously recording and playing back an audio signal. The echo is provided by the delay between the time the signal is recorded by one head and played back by the other, an inch or so away. The reverberation time is affected by the speed at which the recorder is operated. If you feed enough of the playback signal back into the input, one sound will continue to echo and echo until the tape runs out.

We have used the terms reverberation and echo more or less interchangeably. This is not strictly true, however. To quote the NAB Handbook; "Echo is a distinct delay that reproduces the original once or several times over. Reverberation is a continuation of the original sound with no separation between the original sound and the continuation." Using this definition, the spring-type reverberation unit produces true reverberation, whereas the tape recorder system produces echo.

VU meters connected in strategic places in the system are quite useful. You need not use expensive meters in all cases, particularly if you are just interested in whether or not audio is present and its approximate level. Some possible locations are the audio output of a TV tuner, inputs to TV modulators and lecture halls or other areas being fed audio. In some cases it is advantageous to monitor the input signal level to your audio monitors. As stated previously, except for the audio console, VTRs and ATRs (which have built-in VU meters), usually an inexpensive meter is adequate. An expensive super-accurate meter equipped with a selector switch can be connected across selected lines to allow the level to be carefully checked. Remember, a VU meter is accurate only when connected across a 600-ohm line.

SPLITTERS & PADS

Sometimes splitters and pads are necessary in an audio system. A splitter is used to divide an audio signal into several outputs. A 2-way passive (no operating power required) splitter will divide a signal into two outputs, each 3 db down from the input. The splitter can be connected backwards, with the output of either of the inputs down 3 db. If both inputs are used at the same time, the output level will depend upon the phase relationships between the two inputs. Splitters also can be used to divide a signal into more than two outputs. For example, a one-to-four splitter has four outputs, each 7 db down from the input. The splitters are calculated, usually, to operate in 600-ohm lines. However, other impedances can also be used if the splitter is especially designed for that impedance.

A pad is a device to reduce a signal level by a designated amount. They are available in various attenuation values. The most popular are 3, 6, 10, 20 and 40 db.

Splitters and attenuators can be purchased or home built for balanced or unbalanced applications. Tables are available so you can design any type of pad or splitter desired. You can also convert to another impedance with a pad. Such tables are available in many publications.

INTERCOM SYSTEMS

An intercom system providing communication between the director, the studio camera operators and floor manager is essential. Other intercom stations can be located in the camera shading area, both studio and film, the film chain islands, the distribution or program switcher area, the audio console, the lighting director, the audio boom operator, and the VTR area.

Intercom communication systems use a headset or speaker and microphone, or a combination of the two. The camera and floor men need headsets, for obvious reasons, as does anyone else in the studio who needs to talk to or listen to anyone else during a production.

If the film chains and VTRs are in a separate room, it is usually more convenient to talk to that area on a speaker. If the operators need to reply, a few headsets scattered around will be ample; personnel should be free to move about the room unhindered. Usually, distribution area personnel, audio console operator (if at a distance from the production switcher), camera shader and lighting man are best served by headsets. The operator at the program switcher should have a microphone and speaker, as it would be uncomfortable for him to wear a headset for his entire shift. The director might want a mike and speaker, or possibly he would prefer a headset. Both should be provided. If the same information is carried on the speakers and on the headset line, no separate paging switches will be necessary.

The intercom speakers in the studio can be muted when a studio microphone is turned on, but the muting can be overridden during rehearsal so all the mikes needn't be turned off whenever the director wishes to speak to the talent.

The foregoing is just one way to set up an intercom system. If everyone uses headsets the system is much easier to design and install, since no speakers, mikes and power amplifiers are required, except one amplifier, mike and speaker to enable the director to talk to the studio.

Several types of headsets are available. One model has only an earpiece, with no facilities for replying. Another has one earpiece and a microphone on a little boom attached to the earpiece, which positions the mike in front of the wearer's mouth. Both types can have one or two earpieces. If only one is used it carries the intercom audio. If two are used the second can carry intercom audio or program audio. For example, the boom mike operator might want to hear how well his mike is picking up, so he can reposition it if necessary. Even if both earpieces are carrying intercom audio, they help to muffle distracting studio noise, such as a loud orchestra or rock group. One earpiece can be pushed aside if the wearer wishes to hear the voices in the studio. Typical headsets are shown in Fig. 12-5.

Camera tally lights are used to indicate to the talent which camera is on the air; picture monitor tallies remind the director which source is presently on the air, and VTR and film camera tallies make it less likely that an operator will inadvertently stop a VTR or film projector that is on the air. Any piece of equipment can be tallied to any switcher; it needs only to be provided with a suitable indicator and the switcher provided or supplied with tally switches or relays.

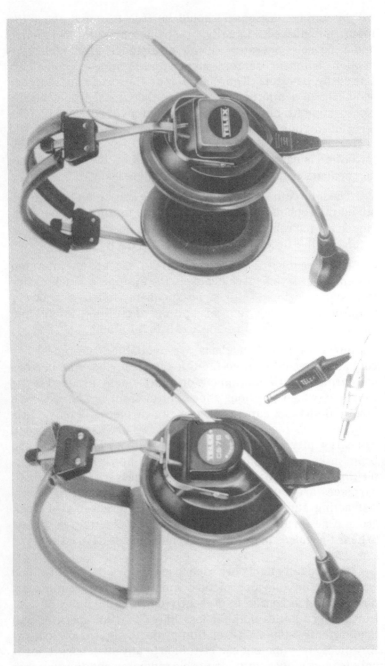

Fig. 12-5. Two popular headsets, both with built-in microphones. (Courtesy Telex)

A production switcher always has a tally light provision; however, the feature is less common on distribution switchers. Almost any switcher can be so provisioned, as long as this is taken into consideration when ordered from the manufacturer. If the switcher model you want does not come so equipped, external tally relays can be added later when the system is installed, as long as one unused pair of normally open contacts are available. The contact current carrying capability is not important, since the contacts can be used to operate a low-voltage relay which will carry the main load.

You will also need a few DC power supplies to operate the tally lights and any external relays. The interphone system sometimes requires a separate power supply, too.

EQUIPMENT RACKS

Fig. 12-6 shows an assortment of equipment racks. Racks come in various sizes and can be provided with side panels, back doors and tops if desired. When more than one is used, they should be bolted together, if mounted side by side. The size used for television accepts 19-inch wide panels. Other widths are also available but are not generally used for television. Shelves are available to mount on the front of the racks; for example, below the distribution switcher to hold the operating log and possibly to mount remote control panels. Blank panels should be inserted in areas not filled with equipment for neatness and to enhance cooling convection currents. You can buy fans that mount in the top or front of the racks to help in cooling the equipment. For exceptional situations, rack size air conditioners are available, too.

Each enclosure should have AC power strips mounted in it to power the equipment in that rack. A duplex outlet mounted on the bottom front of several racks is convenient to furnish power for test equipment, trouble lights, electric drills, soldering guns, vacuum cleaners, or other equipment used in that area.

Another type of enclosure, called a low silhouette rack or console, can be used for the camera shading controls and monitors, and even the production switcher in a small system. The only real advantage these have over a regular rack with a shelf is appearance. Sometimes it is a little easier to service equipment in this type of enclosure; for example, it is low

Fig. 12-6. Two types of equipment racks. Notice the AC receptacles in the one on the left and the facilities for cooling in both units. (Courtesy Emcor)

enough so that you can remove the top and service monitors without removing them.

The audio console and production switcher can be set on or into a table or desk or, if it is large enough, purchased with its own special floor stand.

REMOTE CONTROLS

You will probably want to operate some equipment by remote control. Some possibilities are ATR, CTR, and VTR start and stop, and possibly record. You usually will need not remote the fast forward and rewind functions. You will want to remote control the slide projector on-off and change functions. Remote film projector controls might be forward-reverse, show, start and stop. You may also want to control the multiplexer mirror to select the proper input, such as slide or film projector. Sometimes a fixed prism is used instead of a movable mirror. In this case you will have to be able to douse the film projector so you can run it to cue a film without its light entering the multiplexer. This can be done by turning off the projector lamp or with a movable metal shutter blade to block the light. Of course, if the multiplexer is not being used at the time, dousing is not necessary.

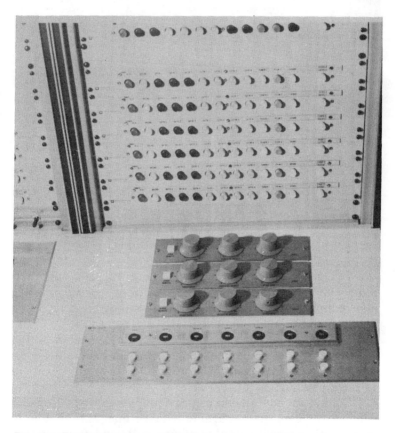

Fig. 12-7. The lower portion of the picture shows a VTR remote control panel. Above that are proc amp remote level controls for quad VTRs. A distribution switcher panel is seen above them.

Fig. 12-8. Remote controls for two film islands are seen in the lower portion of the picture. Above them is an intercom paging control panel. Above that are distribution switcher control panels.

You can put remote controls wherever your operation requires; for instance, at the program, production or distribution switcher or at the camera shading area. ATR and CTR controls might be installed at the audio console.

You can usually purchase remote control panels with the various pieces of equipment. Some of them are rather simple—and expensive enough—so that you might want to fabricate your own.

Equipment control can be provided in two ways. You can connect the controls in parallel if more than one position is used, so all controls are hot at all times. If you are afraid that equipment might be operated inadvertently if set up that way, you can provide a switch at each piece of equipment to delegate control to any area or choose local control. Remote control in areas not selected will not be operative. If you use the delegation system you **must** always remember to delegate control to the proper area ahead of time or you will not be able to operate the equipment when desired. This can be very frustrating when it happens during a production or playback. We prefer to use the parallel system. If you use parallel control you may wish to leave off the record funtion, so it will not be possible to erase a valuable tape if the wrong button is pushed. Record is usually interlocked so that both the play and record buttons have to be pressed at the same time to enable the record function, but still, accidents can happen. A VTR remote control panel is shown in Fig. 12-7 and film island controls in Fig. 12-8.

Chapter 13

The Maintenance Shop

If you are going to do your own equipment maintenance work, you will need an area set aside for the purpose. It should include a work surface, test equipment, spare parts and tools. It can be as elaborate or simple as befits your needs. We will go into some detail about tools, test equipment and maintenance, allowing you to decide what best fits in with your operation and talents.

You will need some test equipment. Some sort of meter is useful. A VOM (volt-ohm-meter) is good for a starter. These can be purchased very inexpensively or a more costly model can be selected. The more expensive meters are usually more accurate and reliable. A unit with active components can be purchased, with a transistor or FET (field-effect transistor) input, which gives more accurate readings due to the higher input impedance. The higher the input impedance, the less effect the meter will have on the circuit being measured; therefore, the reading will be more accurate. Another version of the test meter is the digital type which indicates the measured quantity on a numerical readout instead of a conventional meter movement (Fig. 13-1). This is a nice feature and makes the instrument much easier to read, but it is quite a bit more expensive as this is written. Another meter type is called a VTVM. This instrument has a high input impedance and uses active circuitry. VTVM stands for vacuum tube volt meter.

What with FET inputs and digital readouts, the differences between VOMs and VTVMs is becoming rather blurred. Don't get confused; just pick an instrument that will do what you want at a price you can afford. Look for accuracy, high input impedance, and either AC or battery operation as suits your needs. Also look for a good spread of resistance, AC and DC voltage, and current ranges. If you are working with solid-state equipment, a very low voltage range, 1-volt full scale or so, is very useful.

Incidentally, many of the meters mentioned above, as well as other meters and test equipment discussed in this chapter, are available as kits at a significantly lower price. Heath, Eico, Knight and many others sell such kits.

Another meter often needed is an AC VTVM. This is a very sensitive AC voltmeter with a wide, flat frequency response. In a television system it can be used to check audio levels for system calibration and adjusting audio tape recorders. It can also be used to check the residual noise output of an amplifier, since most will read down to -60 dbm or so. The scales are calibrated directly in volts and dbm.

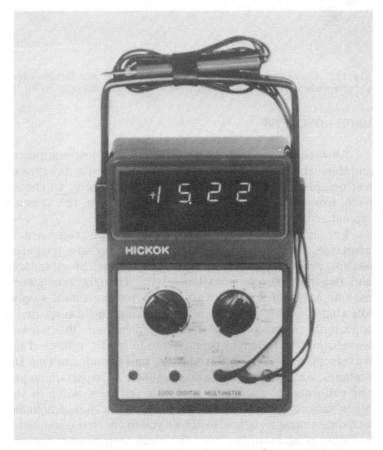

Fig. 13-1. A multimeter with digital readout. (Courtesy Hickok)

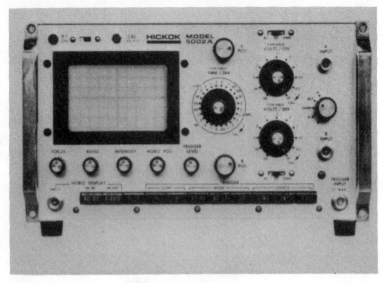

Fig. 13-2. A nice shop scope with DC and AC inputs and facilities for simultaneous display of two signals at once. (Courtesy Hickok)

OSCILLOSCOPES

An oscilloscope is a very useful piece of test equipment and should be included in every test setup. As with any type of test equipment, you have a wide range of quality to choose from, from the inexpensive to the elaborate. Fig. 13-2 shows a typical scope.

For shop troubleshooting, a TV repair type scope, widely available, can be used. Features to look for are triggered sweep (expensive), DC input, accurate vertical attenuator and flat frequency response to 6 MHz. The triggered sweep uses an external waveform to trigger the horizontal sweep, allowing a stable sweep and permitting the sweep to be repeatedly started at a selected point on the viewed waveform. The DC input permits you to see DC offsets of the waveform, view very low-frequency pulses and measure DC voltages. An accurate vertical attenuator is important so you can calibrate your scope on any scale and then switch to any other scale without losing accuracy. The flat frequency response characteristic also allows you to check the frequency response of an amplifier over the video frequency range and view a waveform without distortion added by the scope.

With some scopes you can change the vertical and sometimes the horizontal signal amplifiers and timing circuitry by inserting appropriate plug-in modules. For example, a high-gain vertical amplifier can be fitted to view very low-level waveforms, or a dual- or even 4-trace vertical amplifier can be used to compare several waveforms at the same time. Many other modules are available for almost any requirement. Usually, scopes that accept plug-in modules are rather expensive. An adapter is available that can be connected to any scope to convert it to a multiple-trace capability. This feature can be very useful when troubleshooting.

Usually, with a good multipurpose meter, signal source and scope you can fix almost any problem. Other kinds of test equipment are useful for special tests, and some can save time when testing specific components.

SIGNAL GENERATORS

Very many signal generators are available. The types you might find the most useful are audio, video and audio, RF, and video sweep generators.

An audio generator is used to provide pure audio waveforms at specific frequencies. Sometimes a meter and attenuator are built-in so an exact level can be established. If not, an AC VTVM can be used to set the required level. The output impedance should be 600 ohms. A video test generator, as previously described, supplies color bars (if you need them), multiburst, sine-squared and window, and stairstep waveforms. You will probably use the same generator you use in studio checks.

The RF test generator provides specific RF frequencies for use in adjusting the tuners and IF sections of television sets. It is also used for checking RF distribution systems. The RF sweep generator should be equipped with a marker generator to mark specific frequencies of interest, such as the sound IF frequency and other important points on the TV IF bandpass. A popular model is shown in Fig. 13-3.

Audio and video sweep generators are not used very often because a multiburst signal provides almost the same test signal for video and an audio generator supplying discrete frequencies will do the job in audio servicing. However, sweep generators make the tests easier, faster, and somewhat more accurate.

Fig. 13-3. An RF sweep generator for testing TV sets. A marker generator is built into the unit. (Courtesy Sencore)

POWER SUPPLY

A variable voltage power supply with a meter to monitor voltage and current is very useful. For solid-state equipment it should provide 24v DC at at least an ampere of current. This will cover most potential uses. Neither the positive or negative output should be grounded. Sometimes two such supplies are needed, when both a negative and positive voltage are required at the same time.

COMPONENT TESTS

Next we consider test equipment for specific components. Usually, signal tracing techniques and an ohmmeter will give an accurate enough indication of a component's condition, but to be sure, the following testers can be useful.

Transistors can be tested in and out of a circuit with an ohmmeter. If you are not familiar with the procedure, here is how it is done. Set your ohmmeter to the R X 100 scale (don't use a higher scale) and place one probe on the base of the transistor and the other on the collector. Notice the resistance reading. Then move the probe from the collector to the emitter lead (the other lead should still be on the base) and notice the reading. Both should be about the same and either very high or very low in resistance. Let's assume the reading is high. Now reverse the leads and measure from base to collector and emitter again. The reading should now be lower. Signal transistors will show this characteristic much better than power transistors. Those differences in resistances show the transistor is probably OK. If you read an open or short from base to either collector or emitter both ways, the transistor is probably defective. However, sometimes intermediate ambiguous readings are caused by external components if the transistor is connected in the circuit. In this case you will have to measure the voltage on the various junctions and use signal tracing methods to check the transistor or remove it from the circuit and try again.

Of course, in-circuit transistor testing with the ohmmeter is done with the equipment turned off. It has been said that this method of testing transistors runs the risk of damaging them. But with a meter setting no higher than the R X 100 scale, I have never damaged a transistor testing it this way, and I use the method all the time. This method holds good only for bipolar transistors; I've never tried it on FETs or unijunctions or others, so I can't say if it would work. I doubt that it would.

Special testers can be purchased to check transistors in and out of the circuit more rapidly than with an ohmmeter (sometimes), and usually the result is more decisive. You can go a step further and check the beta (gain) of the transistor under average conditions, or use a curve tracer and check the transistor under various operating conditions. A beta type tester is shown in Fig. 13-4.

You can get a tester to check vacuum tubes, but we feel unless you have a lot of vacuum tube equipment, it is easier to provide yourself with spare tubes of every type used and test by substitution. This method is faster and, unless a lot of vacuum tube gear is used, cheaper. If you use the substitution method, be sure your spare tubes are good by checking them in a circuit. Sometimes new tubes are defective.

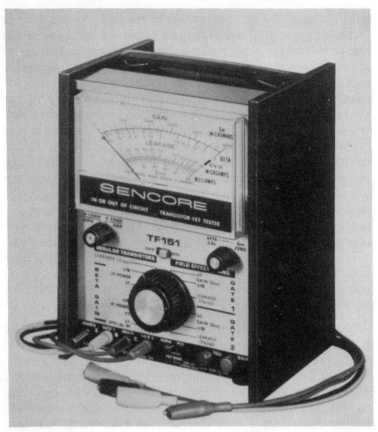

Fig. 13-4. A tester for bipolar and field-effect transistors. (Courtesy Sencore)

Bridges are available to check capacitors and inductors. Usually, an ohmmeter or signal tracer will locate a defective capacitor or coil, but in inconclusive cases a bridge will assure you of the correct diagnosis. In this case, of course, parts substitution will also back up your diagnosis, if the correct part is available.

A useful piece of test equipment is the CRT tester-rejuvenator (Fig. 13-5). To go with this, lay in a supply of tube brighteners for the pictures tubes in your monitors. You can use a brightener until it no longer works, then try the rejuvenator. This can add quite a few months of life to CRTs. The tester portion is used to check a CRT suspected to be defective.

SPARE PARTS

The spare parts you should stock for your system obviously depend on the equipment you are using. A few general recommendations follow.

A good selection of resistors will come in handy. You should have a few of every 10 percent value of ½-watt rating. Every 5 percent value is better, but it requires many more resistors. You can replace higher wattage resistors by paralleling ½ watt units. Remember, use identical values in parallel and divide the value of one of them by the number of resistors in parallel to obtain the resulting resistance. For example, three ½-watt 150,000-ohm resistors in parallel equal one 1½-watt 50,000-ohm resistor. Of course, if your budget permits, you can stock the higher wattage resistors as well.

Fig. 13-5. A color or monochrome CRT tester-rejuvenator. (Courtesy Sencore)

This allows you to make neater repairs, as well as eliminate possible problems. In critical RF circuits two resistors in parallel will provide the proper DC resistance, but the added inductance is liable to upset the circuit. It might be difficult, too, to find the room for additional resistors.

It is not an easy matter to suggest what to stock in the way of spare capacitors. Electrolytics are the most prone to fail, so some of the values most used in your equipment might be stocked. Usually, a capacitor up to 100 percent larger in value can be substituted for electrolytics in most circuits if you don't have the proper value. This does not hold true for capacitors in resonant or timing circuits. And in modern miniature equipment the capacitor usually has to be the right size or it won't fit in the allotted space. Mica and tantalum capacitors seldom fail unless the equipment is very old. Paper capacitors do fail, as do mylar capacitors, so a few common values of those types should also be stocked.

A few yokes and flyback transformers to fit your monitors and TV sets should be on hand, as they fail from time to time and are best tested by substitution. A few spare width coils are also useful.

Spare CRTs and other vacuum tubes should be kept on hand, especially horizontal output tubes and dampers for your monitors. In our experience these seem to be the most likely to fail. We find it best to replace both if either one fails.

Transistors and diodes should be stocked, too. But the cost of a few spares of every type you have in your equipment can be prohibitively expensive. A better start can be made by stocking a set of the universal replacement transistors and diodes manufactured by GE, RCA, Motorola or others. They supply cross-reference charts listing substitutes for most of the JEDEC numbers (1N—or 2N—). If you find that a transistor has failed which cannot be replaced by one of the universal replacements, you can purchase a few of that type, using one to replace the defective unit and the others for stock. Eventually, you will build up a supply of semiconductors that will replace any that fail.

A good selection of fuses is essential, both regular and slow-blow.

Some specialized parts will have to be stocked, such as parts for your VRTs. Spare VTR video heads should be kept on hand, as well as belts and drive wheels. You should have a tip projection gauge to properly position a new video head in your VTR or to adjust worn heads. The proper procedure should be outlined in the manual. If you do your own maintenance, be sure to get shop manuals for your VTRs and other recorders, if they are not supplied with the equipment. This means shop manuals, not operating manuals, though they are useful, too. Shop manuals can be invaluable for maintenance and purchasing spare parts.

Other equipment uses parts that will require frequent replacement. For example, belts for film projectors, projection lamps, pilot lamps, etc. You know which parts will probably give you trouble; if you don't, you will before very long.

TOOLS

There is an almost infinite collection of tools you can buy for your shop. We'll name just a few of the basic ones. You can buy others when you find need for them.

Small, medium and large screwdrivers, philips and standard
  Screw-holding screwdriver with a long shaft
  Needlenose pliers
  Diagonal cutters
  Hemostats for heat sinking when soldering or unsoldering
  Soldering gun, dual-heat type with extra tips
  Soldering iron, low wattage for delicate work
  Solder removing tool or braid
  Wire stripper
  Electric drill and bits
  Set of TV alignment tools
  Set of nutdrivers
  Adjustable end wrench
  Contact burnishing tool
  Allen wrench set
  Spline wrench set
  Pocket knife
  Magnifying glass

You'll probably think of many others as you go along.

MAINTENANCE TIPS

Please don't jump right in and tear down a piece of equipment you think is malfunctioning without thinking the situation over for a minute. Maybe there's nothing really wrong. You might have something patched wrong or the equipment might be turned off. For example, I won't tell you how long it took me to fix a viewfinder one day when all that was wrong was that it was switched to external input. Another time I spent hours troubleshooting a quad tape machine that showed no audio output on the VU meter. It turned out someone had a patch in the audio output patch panel jack plugged halfway in, which effectively shorted the audio output. See what I mean?

After you're sure the equipment is hooked up right, check the fuse. If it has failed, carefully examine it. If the ends of the fuse element are smooth and it is just parted, replace it. If you see fuse material splattered on the inside of the fuse tube and jagged edges on the parted ends, there probably is a short in the equipment which will have to be repaired. Never never use a replacement larger than the original fuse.

Next, check to be sure the operating controls are set about where they usually are. See if there is really an input signal at the equipment; connectors and cables do fail. Check the pulse drives, if used. Look at the output directly with a scope to be sure your monitoring equipment hasn't failed. If it's a camera be sure it isn't capped, that the iris is open and the focus is about right. If none of these checks turn up anything, you might as well pull the equipment and put it on the bench, if it isn't too big. Don't pull it half out of the rack and try to work on it right there unless you must. You'll usually end up running around finding tools and test equipment. You'll also find that usually someone wants to use the equipment right then and will continually be in your way.

To make your job a little easier, run sync, video and audio cables from the patch panel to the shop. Then you can patch in any inputs required for equipment you are working on.

Sometimes you'll have to work on the equipment right where it's installed, such as a switcher or audio console. Then you'll have to use extender boards to get at both sides of the defective module.

If you're working on a TV set or monitor, there are dozens of good books written about TV repair so we won't go into any detail on that subject.

First check the power supply. If the output is low, disconnect the load and check it again. If it's still bad, check the diodes and the filter capacitors. If they're OK, you have a bad transformer (unlikely) or a bad regulator circuit. Try the series or pass transistor first, then you're on your own. If the supply works when disconnected from the load you have troubles beyond that point, probably a shorted capacitor, tube or transistor.

If the DC operating voltages are OK, check the signal path about half way through the equipment. If it's OK to that point, divide the remainder in half and check again. Using this method you will soon, with any luck, have the problem pinned down to a single stage. (Watch out for feedback circuits which can confuse things.) Then check the voltages in that stage to isolate the bad component. If this fails, disconnect and check each component until you find the one that failed. You need only disconnect one end of the component. (It will usually be the one you check last!)

Once you find and replace the defective component, analyze the situation to see why it failed, if possible. Possibly some other component has failed, too, or changed value and caused the failure of the part you replaced. This is worth checking, otherwise the equipment will soon fail again. Incidentally, fuses fail frequently without indicating trouble in the equipment. A line voltage surge is usually the cause. If you tape a spare fuse of the proper type and value next to each fuse holder, you will save a lot of time hunting the right one when you are in a hurry.

When a helical VTR gives you trouble, suspect a dirty machine and head first, then mechanical problems. A worn head will require optimization and equalization adjustments; the trouble symptoms resemble signal system or servo problems. Also check for wrinkles, creases and ruffled edges of the tape. This can cause all sorts of trouble. Also check the interchange adjustments.

If you have two pieces of modular equipment of the same model, you can usually pin a problem down to one module by swapping them one at a time with boards taken from the good equipment. This can save a lot of time. If you can afford them, spare modules for one-of-a-kind equipment that is essential to your operation will prove to be well worth the expense.

Quad VTR maintenance is an art in itself. If the reference, tach, capstan discriminator and the drum and capstan oscillators are set correctly, you may have problems if the machine is not operating properly. Also be sure to check the equalization and the standards switch setting. Also try replacing the video head assembly. Be sure the tape you are trying to play back is a good recording (play a tape you know is good). Also be sure the machine is getting reference sync.

Also get a standard alignment tape for your helical VTRs, so the tape guides can be adjusted for proper interchange. With the correct adjustments a tape recorded on any VTR can be played back on any other properly adjusted VTR. If the VTRs are not adjusted for proper interchange, it might be impossible to get a tape to track properly. Some machines require frequent checks and adjustments of the interchange capability; others are not prone to wear. If the guides are misadjusted or worn, it will show up as a critical tracking adjustment or it might be impossible to eliminate the tracking error in some portion of the picture.

The tracking control on a VTR adjusts the drum speed to position the head in the center of the band of recorded information on the tape. If the head leaves this area, noise will result because the head is no longer playing back the recorded material. If greatly misadjusted, it might even play back part of the material on an adjacent track. The tape guides position the tape around the drum. If one of the guides is worn or misadjusted, the head will leave the recorded material at that point. The alignment tape is a standard tape all recorders should track properly. If all your machines will track this tape properly, they will track tapes recorded on other machines. Your VTR factory manual will tell you how to adjust the interchange on your machine.

A quick way to check interchange is to put an interchange tape or one you know is correct (interchange tape preferred) on the VTR and play it. Slowly adjust the tracking control and if the entire picture starts showing tracking error at the same time, the interchange settings are probably OK. If only one portion of the picture shows tracking error as you slowly misadjust the tracking, you should adjust the interchange, but you can get by for a short time. However, you shouldn't record on this machine. If you can't eliminate the tracking error in some part of the picture, the machine must be adjusted before recording is attempted.

Your shop manual shows how to hook up a scope to check and adjust the tracking, but the foregoing check will quickly tell you if you have problems. Guide adjustment is tricky and you should be sure to check your manual for proper procedures. After some practice you will find it to be a simple adjustment. Be sure when adjusting the entrance and exit guides you don't adjust them in so far that they touch the rotating head. If you do, you will probably destroy one or both heads. Don't get carried away when adjusting the guides! Keep track of how far you turn each adjustment. If it doesn't produce the desired result, return it to its previous position and consult the manual again. If you get all the adjustments way out, you will have some trouble getting them all back again.

There are several methods for cleaning tape recorders. You can clean the metal with Freon TF and the rubber rollers with alcohol. We use Miller-Stephenson MS-200 spray to clean everything. Check your manual for permissable fluids and cleaning procedures. The MS-200 spray is kind of expensive, but is very easy to use. We find that since it is so easy to apply, the tape recorders get cleaned more often, resulting in better operation. VTRs especially need to be cleaned often, every time they are used is not too frequent. But if the operator has to find two bottles of cleaning fluid and Q-tips, and he has to clean the machine very meticulously, he is liable to skip the procedure frequently. On the other hand, if all he has to do is pick up a spray can and spray the contact surfaces of the VTR, he will do so more readily. Thus, the added expense of the spray cleaner is offset by fewer head clogs, less tape drag and other problems.

If you loose the audio output from the film projector, check the exciter lamp first. If it's OK, check the oscillator tube that supplies its voltage.

Be sure to check the levels (audio, video and pulse) in your system frequently. This takes time but can forestall problems.

If any of your equipment uses air filters, be sure to clean them frequently so they can do their job.

Chapter 14
Putting A System Together

When you have reached agreement on the design of a system, you might, in some cases, want to put part or all of it out to bid. This could be required by the school system or company you are associated with, or you might feel this is the best way to get the lowest price for the system you desire. Obviously, the following information will not hold true for every situation. Just choose that information applicable to your needs.

SPECIFICATIONS

Briefly, a bid specification should set the minimum quality of the equipment required and help you get a good, workman-like installation job. It should also guarantee an acceptable level of expertise of the installers.

A specification is usually divided into several portions. We will assume that it is to be a "turnkey" system; that is, one that is completely finished, tested and turned over to you ready for operation.

We can start with the "boilerplate" specification, a term sometimes used to describe the part of the specification that sets forth some or all of the following requirements. It can call for an outline of the qualifications of the bidder; that is, the length of time he's been in the systems installing business, the number of other comparable systems installed, together with their locations and time in use, spare parts facilities and experienced technicians available for warranty and later service as required. The warranty time you require should also be stated.

You can outline the time the areas under construction will be accessible to the installers and the availability of space for storage of equipment and tools during the installation period, including the security, if any. You can present a detailed description of the required testing of the system and who will accept it, or make the submission of the test procedure part of the bid. It would be preferable to draw up your own test procedure, making sure that all parameters are adequately tested. In either case, this would be a good time to lay down the technical specifications of all the equipment, connected together and operating as a system. The manuals, drawings, cable run sheets, etc., required can also be outlined. Don't forget to stipulate that any damage caused by the installers will be repaired.

You could also ask for on-site training in the operation and maintenance of the system. You should also ask for a guaranteed installation date. It would be well to add a paragraph requiring communications with the local chief engineer concerning technical coordination; i.e., equipment location details and other such subjective information. The above are only a few of the items you can include that are not covered elsewhere in your specification.

You could then go into what might be termed your design philosophy. This could include a description of what you expect the system to do for you—how it should work together. This can be of help to the bidder, allowing him to visualize it as a system, rather than a list of equipment and some block diagrams.

You should supply the following flow diagrams, in block form; video, audio, pulse, control, tally and intercom. The video diagram should include all the equipment hardwired into the system—major equipment, patches, terminations, unused inputs and outputs, sync generator remote inputs and crosshatch outputs, and peripheral equipment. You should show whether a piece of equipment is to utilize a looping input or should be terminated. Don't forget to show if a picture monitor is to be looped or fed by the waveform monitor video output jack to utilize the single-line monitoring feature. Any balanced video lines should be clearly differentiated from the normal 75-ohm unbalanced circuitry.

The audio diagram should be similar to the video, showing all terminations, patches, etc.

The control diagram is not absolutely necessary. You could just specify the functions of each piece of equipment that must be remote controlled—video tape recorders, audio tape equipment, film islands, video and audio levels, etc. Specify each function to be controlled and each point from which they are to be controlled. Don't forget to specify parallel or delegated control. Some equipment may require rewiring to use parallel control. Distribution switcher remote switching can be specified in the switcher specifications.

A power supply with good voltage regulation over a wide range of current should be specified for the tally light system. If you use a distribution switcher, you might only want tally lights to work if certain output buses are used. This can usually be incorporated into the switcher specification. Don't forget to specify the placement of tally lights on equipment that normally does not have them, such as monitors.

Describe the type of intercom system you desire. It should be outlined in detail, together with a drawing if necessary. In any case, be sure to detail all interphone jack locations. Don't forget muting of the studio intercom speakers when a mike is turned on. Also include the number of speakers and headsets required.

You should try to lay the equipment out in the number of racks you plan to use to ensure enough rack space. If you have enough racks, you should allow as much room as possible for future expansion. You would be surprised at how far you can spread out in a year or two.

To help the bidder estimate the cable required, a floor plan of the area should be supplied. It should give the dimensions of each room and the relationship of the rooms to each other. Be sure to show the scale of each drawing.

It would be wise to submit drawings of the audio, video and pulse patch panels to show where each patch in the system is to be located in the jackfield, allowing the most convenient placement for your needs. Again, be sure to allow sufficient room to spread out later.

Power wiring can be considered next. Specify the number of power strips in each rack and which, if any, will be paralleled and fed from one power panel breaker. In a turnkey installation you might save money by specifying that you will supply the power from the breaker panel to the contractor supplied and installed power strips. While on the subject of power, don't forget to specify auxiliary outlets on the bottom of some of the rack fronts to power test equipment, etc. Another point to keep in mind is to try and balance the load evenly between the three phases of power in your panel.

Regarding equipment, you have a choice of several ways of writing the specifications. You can copy the specification sheet of the model you want, or you can just name it and specify that unit or an equal. However, you may get some static about what is equal to what if you do it that way. A third way, if you have no particular piece of equipment in mind or want it to be a custom model, is to list the physical and electronic specifications you require and let the contractor supply whatever will meet your specifications. Be sure the specifications are not so loose as to cause the system to fail to meet the required performance specifications or so tight as to be priced out of sight.

Next you could consider outlining construction practices. Leave as little as possible to chance and you will be less likely to be disappointed. For example, you might consider some of the following items: wire markers on the cables, cable run sheets, cable lacing, separation of cables carrying different signal levels or types of signals, methods of labeling patch panels, finish of contractor supplied control panels, etc., mounting film chains, wall mounting of speakers and picture monitors, grounding, cable and connector types, and other details not spelled out elsewhere in the specification.

In general, wherever there might be any doubt as to what you want, don't be afraid to go into some detail. For example, don't just specify a film island to be installed in a certain room; show just where you want it and the picture monitor on a scaled floor plan, and specify just how you want the mounting area leveled and how the units are to be fastened to the floor.

You could now make up an equipment list, by model numbers or specification paragraph number. Be sure to number every paragraph or sub-section of the specification for reference purposes. Designate quantity or "as required" for items like blank panels, cables, terminations, connectors, etc. List everything you need, including things like rack side panels and work-writing shelves, extra rack hardware, matching paint, headsets, tables or consoles for the audio console and video production switcher, shelves for mounting monitors, cable tray, test equipment, spare boards, extender boards, black generator, patch panels, interphone system power supply, camera cable patch cords, camera cables, camera cable patch panel and studio panels, mike and interphone jacks in the studio and at the lighting panel, floor monitor outputs in the studios, terminal blocks, monitor and intercom speakers, projector and camera lenses, mikes, test pattern illuminator, video test generator, and any other items you may have overlooked. If you can afford it and don't have the maintenance time and-or personnel available, order spare boards, monitors, power supplies, distribution amplifiers, and anything else you can think of. This can save a lot of downtime.

Just to be on the safe side, it is a good idea to put a paragraph in the specifications somewhere that says in effect that certain items such as miscellaneous hardware, surface conduit, connectors, cable tray sections, etc., though not specifically mentioned, shall be furnished and installed by the contractor, thus insuring a complete, operational system. This covers you if you leave out some relatively minor items. You will probably find specification writing interesting but tedious. It will undoubtedly take a lot longer than you had anticipated. The biggest problem is finishing; everytime you reread it you will find something to add or change.

SYSTEM DESIGN

Some of the material included here was covered in part earlier, but it will be treated again for the sake of complete coverage. The best first step is to get the system down on paper, with a nice big eraser handy. Design each system separately, video on one sheet, audio on another and pulse on the third. If you want to, you can also lay out a diagram of your remote control and intercom systems. A few of the common symbols appear in Fig. 14-1. Other pieces of equipment are shown as blocks or rectangles and labeled what they are. The layouts and symbols outlined are commonly used, but some designers use others. Feel free to lay the system out in any manner that suits you.

For the first drawing, you can use pen or pencil, and a drawing board and drafting tools, or draw freehand. Most designers do the first draft freehand. If the final draft is to be part of a bid specification, it should be done in ink so the original will not be smudged and will reproduce well. Of course, bid specification drawings should be done neatly, using a T square, triangle and templates, if only for your own desire to do a professional looking job on drawings that will be seen by so many people.

Let's start with a video system. Lay out the sources you have decided on in a vertical row on the left side of the sheet.

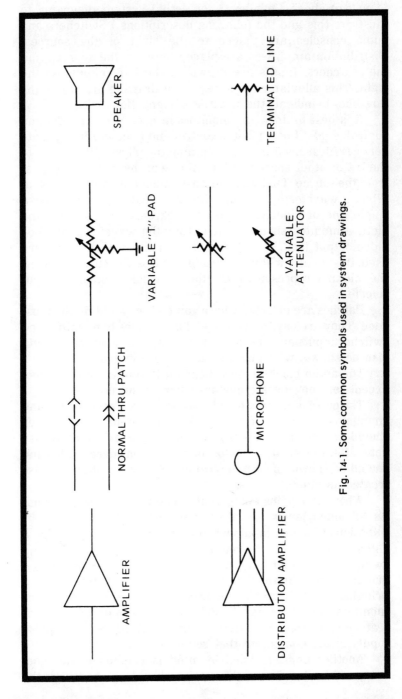

Fig. 14-1. Some common symbols used in system drawings.

(You can check Chapter 15 for actual system diagrams to illustrate this and the following descriptions.) Switchers and audio consoles are placed to the right of the sources. Distribution amplifiers are placed between the sources and the switchers. Inputs are drawn on the left, outputs on the right. This alleviates the necessity of drawing arrows on the flow lines to indicate the direction of signal flow.

It is best to draw the monitors in each area (production control, studio floor, master control and telecine) in separate vertical files. Loop through the monitor physically closest to the DA or other source and terminate at the monitor farthest from the source. That way the least amount of cable is needed. When a waveform monitor with the capability of viewing a single line of video, such as the Tekronix 529, is used, the picture monitor should be fed from the waveform monitor video output jack. If both A and B channels of the CRO are used, this feature will not be usable, unless you only need view the picture represented by the waveform monitor input selected.

Patches are inserted where you feel they will do the most good. They are particularly useful in the lines to monitors and switcher inputs and outputs. Unused input and outputs should also be shown, with unused outputs terminated.

The audio block or flow diagram is similar to the video, except for showing attenuators where required.

The pulse diagram should show the sync generator(s) and changeover switch (if used). All equipment using pulses should be designated, with the type of pulses required also noted. Remote sync generator inputs are considered video and should be shown on both diagrams. Crosshatch output is also treated as video.

When interfacing equipment, video should be no problem. As all video levels are alike, you need only note the termination of looping inputs when they are not used to bridge a signal. Remember, if you are using a passive bridging switcher, its output must feed a very close by unterminated bridging input. Be sure the equipment you select indeed has a bridging input, if that is required. Also be sure the picture monitors you select can be driven by external sync, if that feature is desired. Be sure you supply all the required pulse inputs to the equipment that requires them.

Another point to keep in mind is composite and non-composite video. The right version should be used with the right equipment, of course. This requirement varies with the equipment selected. Check the specification sheets, which give this information.

It is also a good idea to lay the equipment out to scale in the equipment racks to be used. Pulse, video and audio equipment is usually mounted in separate racks, if numerous enough. This is not necessary though. Audio cables should be separated from pulse, power and video cables. Audio cables carrying different levels should also be kept away from each other. In general, keep the different cables in different laced bundles to avoid possible interference.

If enough space is available, put a 1¾-inch blank panel between equipment. This allows freer circulation of air and aids cooling. It also makes it easier to mount and remove equipment from the rack for repair.

Cables carrying the same type of signals should be neatly laced together and routed to the sides of the racks. If cables are allowed to hang loose in the racks, you will soon have an unmanageable mess, and access to the rear of the equipment will become very difficult. Also allow some slack in the cables connected to the equipment so it can be partially removed from the rack for service without having to disconnect all the cables and make up jumpers.

While we are on the subject of cables, they should all be numbered, with the same number on each end. A prefix letter can be appended to each number, or different colors can be used to differentiate between pulse, video, RF and audio cables. Each cable number should be logged, with information as to where each end is connected (equipment number, type and function).

If you have more than one of any piece of equipment, each unit should be numbered on the front panel to keep them separate for maintenance records, cable run sheets and block diagram designations. If decals are used they should be sprayed with Krylon clear spray or a similar product to protect them from being scraped off.

The patch panels should be laid out in some logical order so you can easily find the jack you want when you are in a hurry . It is also a good idea to color code the video to differentiate between composite and noncomposite video. It can be very disconcerting to patch in the wrong type when you are in a big hurry and definitely do not need the wrong type of video right at that moment.

When planning the primary AC wiring, put each rack or group of logically associated racks on a separate circuit breaker in the power panel. Use two power strips in each rack and separate equipment that can remain on 24 hours a day from that which is only used from time to time. Picture and waveform monitors should be turned on only when needed. Each power strip is then connected to a different breaker. This allows groups of equipment to be turned on and off at the breaker panel without having to turn each individual piece of equipment on and off. It also makes it less likely that a monitor or other piece of equipment will be left on inadvertently. Try to divide the load between the three phases of primary power available at the breaker panel.

Grounding your system properly is very important. Use heavy braid or, better yet, a thin copper strap to ground the equipment racks together and to the non-rackmounted equipment. Poor grounding will show up as a buzz, hum or poor signal-to-noise ratio in the audio and a ripple in the base line of the video, when viewed at a field rate, or an upset of the vertical interval. Clamping the video signal will remove this, but it is best to eliminate the problem to begin with.

Audio system grounding is a little less straightforward, especially mike lines. For example, you might try grounding all the shields at the audio console and letting them float at the equipment. Then hook up the shields one at a time and see if the residual noise level increases or decreases. High-level audio grounding is not so critical and usually both ends of the shield should be connected to ground. If hum is evident you can try disconnecting one end or the other.

Audio cable is available with braid or foil shielding and solid or stranded conductors. Foil shielding is the easiest to work with because it has a solid conductor wrapped in the shield to make the ground connection, thus eliminating the tedious unbraiding and twisting of the braid to form a conductor to make connections. Solid conductors are easier to work with, but tend to break if frequently flexed. It is best to use solderless ring connectors or spade lugs for screw type audio connections. You could loop the conductors around the screws, but they tend to work loose and sometimes short to adjacent screws.

BNC and UHF connectors can be soldered to the ends of video or pulse cables or crimp-on type connectors can be used.

(Video cables are discussed in an earlier chapter.) The crimp-on type are far easier to use, but a special tool for each type of connector and cable is required. The tools are rather expensive, but if a lot of connectors are to be used the time saved will make up for the additional expense.

Scaled drawings of the rooms showing the placement of racks and other equipment should be made to be sure there will be enough space and that there will be enough room around the equipment to operate it. For example, the film islands could take up more room than you anticipate, so be sure you can get around them all right and that they don't intrude on space required to maintain or operate other equipment. Fasten the film island components firmly to the floor, and be sure they are leveled. Also be sure to choose the correct focal length lens for the film and slide projectors to work with the multiplexer you choose.

The routing of cables from point to point permits several alternatives. If you are moving into or redesigning an existing facility your choices are narrowed. Probably the most versatile arrangement is the use of computer floor. In case you are not familiar with it, this type of floor comes in small squares, about two feet on a side. They mount in a framework about eight inches above the supporting floor. Each square can be lifted, using a tool containing several suction cups. This allows access to the floor area beneath them. Thus, you can mount equipment anywhere on such a floor and route the cables from place to place without their being in your way on the surface of the floor. The room has to be designed for this floor, so that the computer floor will be at the same level a normal floor would have been. Thus the door sills and power outlets are about eight inches above the supporting floor.

An alternative to computer floor is the use of cable trenches. These also keep the cable off the floor and out of your way. The disadvantage is the fact that the equipment has to be located adjacent to the cable trench, allowing less freedom of equipment placement in such a room.

If all the equipment can be located in a single row, the racks and enclosures can be set on a low platform and the interconnecting cables can be run in this platform. This method allows an ordinary floor to be used. However, some cabling, such as that going to another room and the wiring to the power panel will have to be located on the floor surface away from the equipment. If the arrangement is planned carefully these cable entry and exit points will be only a short distance from a wall and located behind equipment racks. This will lessen the possibility of tripping over them every time you move around the room.

Other less elegant solutions include using conduit, ducts, or exposed cable trays. If high racks are used, the open cable tray can be located above the ceiling (if it has removable panels), and cables dropped via conduit to each rack. Because some racks will require a large number of cables, and since power, pulse, video and audio cables shouldn't be intermixed in one conduit, this can be a pretty clumsy arrangement unless very carefully planned beforehand. Duct or cable tray laid directly from the top of one rack to the next can also be used. However, if the racks are any distance apart, this will result in a rather poor appearance. Cable tray, either in the ceiling or hung just below it, is commonly used to carry the cables from room to room if conduits or trenches under the floor are not provided.

One tip you may find useful where you have to run some cable through a long piece of conduit with several bends in it and in which no pull wire is installed. If you can't get an electrician's snake through it don't despair. Get a large industrial vacuum cleaner and insert the vacuum hose in one end of the conduit without turning it on. Seal it into the conduit using putty, clay or wet rags. Then go to the other end of the conduit and fill a small plastic bag about a third full or air. Use two bags, one inside the other. Seal the bag with a piece of kite string or similar strong cord tied around the top. Now lay out enough string, carefully, so it will not tangle when pulled rapidly into the pipe, to reach the vacuum cleaner. Put the bag into the conduit and turn on the vacuum. If all is well the bag will be in the vacuum cleaner, pulling the string with it, before you know what happens. Sometimes you will have to pull the bag back a bit if it gets stuck somewhere along the way. Then you can use the string to pull in something strong enough to pull in the cables. You will be amazed how well this works.

Chapter 15

Three System Designs

A single VTR—camera—monitor system needs little comment. From the descriptions of the various equipment covered earlier, you can choose one that will suit your needs best. Such systems are usually purchased as a package, with all the components designed to work together. You could choose the conventional VTR, camera and monitor or a portable battery-operated system with hand-held camera. The VTR might only be capable of recording, in which case the tape will have to be played back on another VTR at the studio. In any case, systems like this are mostly a matter of choosing what you need. They require a minimum of design, since that is done for you by the system manufacturer. The units will interconnect only one way. Two examples are shown in Figs. 15-1 and 15-2.

SYSTEM 1

As an example of system design, we are going to put together a small television system, with details of why we did what we did as we go along. This system is illustrated in Figs. 15-3 through 15-7.

Let's start out by deciding what we want the system to do. We'll assume it is a closed-circuit system for a school, with facilities to record programs off the air and from a studio. We want a studio floor monitor which can be used as a program monitor during production or view tapes after production. We'll use two monochrome cameras in our studio, along with facilities for two microphones. The system will have a film chain with one 16mm projector and a slide projector. We will also use two helical scan VTRs, one with editing capability, a small video production and distribution switcher and a sync generator. Some sort of intercom system and tally light system will also be needed. Two RF distribution channels will feed programming to designated areas.

Fig. 15-1. A typical single camera-VTR-monitor system shown in use. (Courtesy Ampex)

Let's start by laying out a floor diagram of our system. The physical layout shown in Fig. 15-4 is condensed to allow two people to easily operate the entire system, or even one person in a pinch. The video shading and production control positions are side by side, so one person can operate both if necessary. Monitors are shared to save expense. The test monitor above the program monitor is supplied by one output of the 6 x 6 distribution system. During a production, one of the VTRs can be switched up on this monitor, if it is to be used in the production. If the director wishes, he can also switch up the camera not on the air on the preview monitor to be more easily viewed at his position at rack 3.

The Cam 2-Film CRO can be used to shade the film chain camera, using the CCU (camera control unit) at rack 1, and the picture can be viewed on the film picture monitor. Normally, each camera would have its own CRO and would be operated with the CCUs in rack 2. If necessary, any input to the distribution switcher can be viewed on the Cam 1-Test CRO and test picture monitor. The CROs have two inputs,

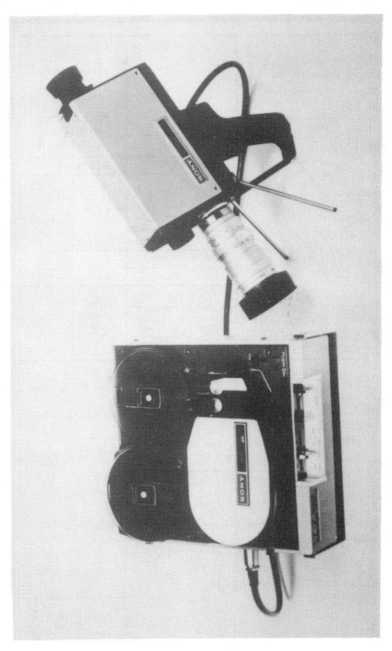

Fig. 15-2. A portable ½-inch VTR and camera. This unit is battery-powered. The VTR can be carried on a strap over the shoulder and with the hand-held camera makes up a very flexible system. (Courtesy Sony)

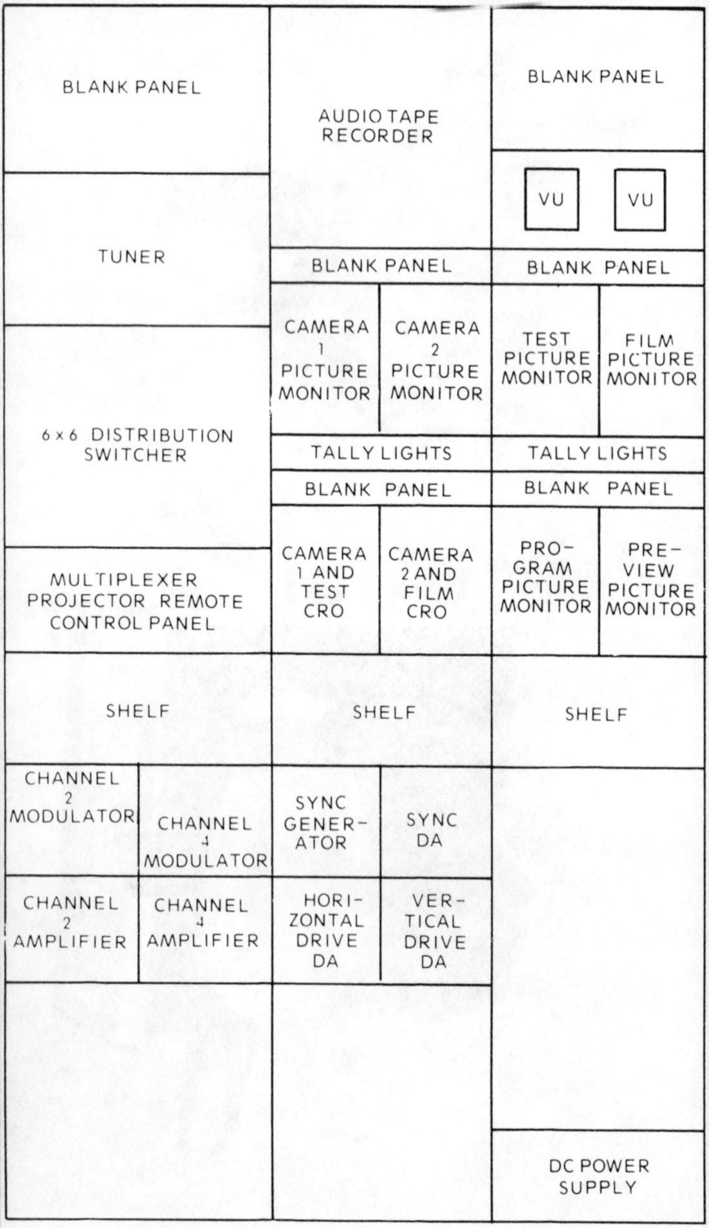

BLANK PANEL	AUDIO TAPE RECORDER	BLANK PANEL	
		VU	VU
TUNER	BLANK PANEL	BLANK PANEL	
	CAMERA 1 PICTURE MONITOR	CAMERA 2 PICTURE MONITOR	TEST PICTURE MONITOR	FILM PICTURE MONITOR
6 × 6 DISTRIBUTION SWITCHER	TALLY LIGHTS	TALLY LIGHTS	
	BLANK PANEL	BLANK PANEL	
MULTIPLEXER PROJECTOR REMOTE CONTROL PANEL	CAMERA 1 AND TEST CRO	CAMERA 2 AND FILM CRO	PROGRAM PICTURE MONITOR	PREVIEW PICTURE MONITOR
SHELF	SHELF	SHELF	
CHANNEL 2 MODULATOR	CHANNEL 4 MODULATOR	SYNC GENERATOR	SYNC DA
CHANNEL 2 AMPLIFIER	CHANNEL 4 AMPLIFIER	HORIZONTAL DRIVE DA	VERTICAL DRIVE DA
			DC POWER SUPPLY

Fig. 15-3. Drawing of the equipment racks for System 1.

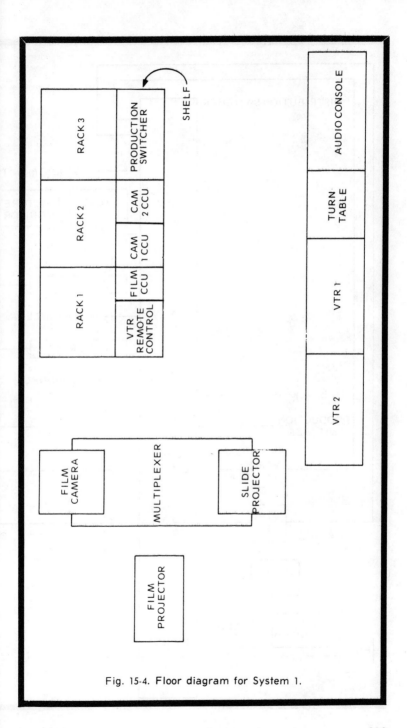

Fig. 15-4. Floor diagram for System 1.

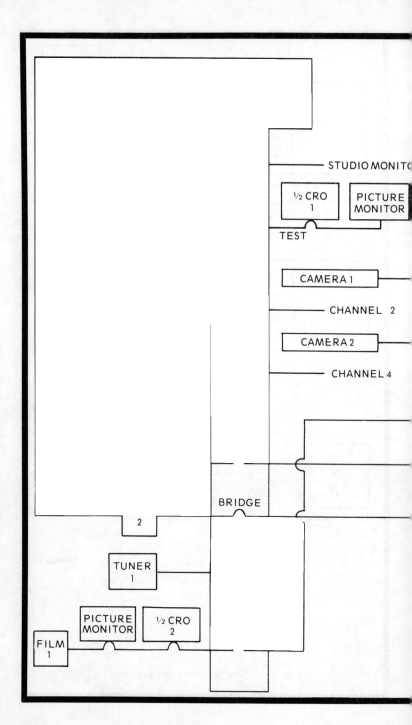

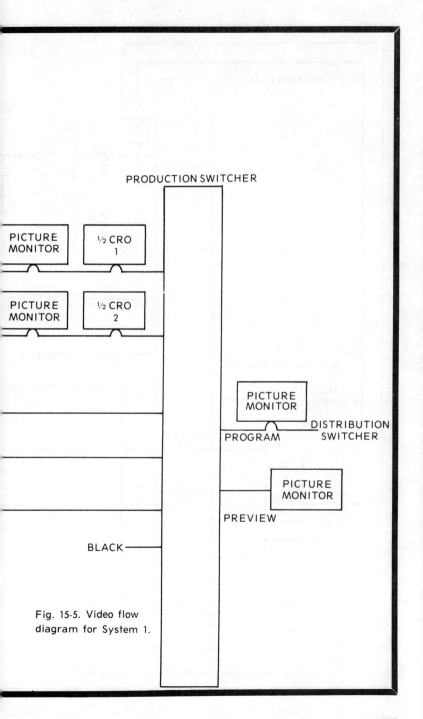

Fig. 15-5. Video flow diagram for System 1.

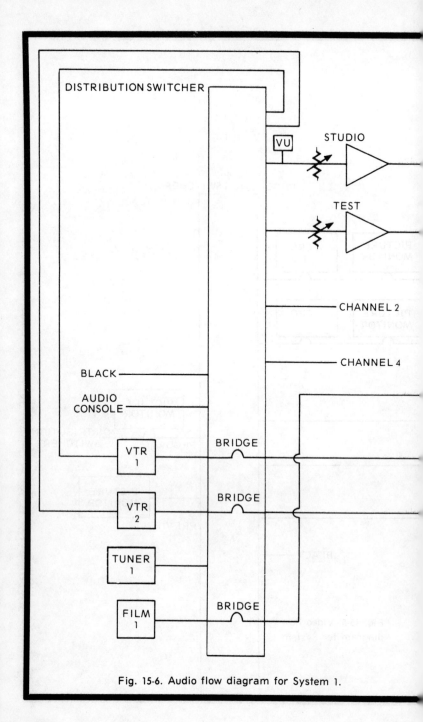

Fig. 15-6. Audio flow diagram for System 1.

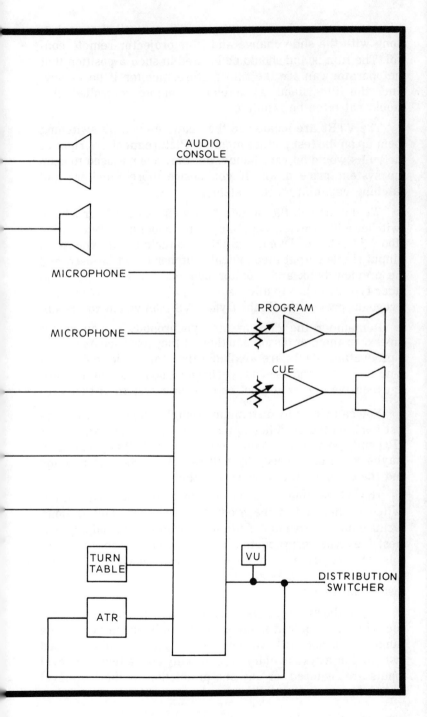

selected by a switch. The multiplexer is controlled at rack 1, along with the slide change and film projector remote control. The film island should be located in such a position that the operator can see the film picture monitor if necessary. Since the film island is completely remote controlled, this should not often be required.

The VTRs are located so they can be cued by switching them up on the test picture monitor. Picture monitors right at the VTRs would be very handy, but they are not used to keep the system price down. If you desire more monitors and patching capability, see System 2.

We did stretch the price a bit by choosing a 2- or 3-bank switcher with preview capability. This could be a Telemation Model TPS-8X2 or Dynair VS-121B. To further cut expenses a 6-input single-output mechanical switcher could be used and the preview bank and monitor deleted. I would advise the fader type switcher to allow supers and dissolves. If you can forego the preview bank, the Dynair VS-150A would do the job.

Remember, the VTRs are nonsynchronous, so you cannot super over them or dissolve to them if they are used as part of a production. VTRs are available that can be locked to the sync generator and used as synchronous sources, but they are so expensive they would not be used in a system of this size.

You can view any distribution switcher input by using the test bank on the switcher to put them up on the Cam 1-Test CRO and test monitor. You can view both VTRs operating in playback simultaneously by putting one on the test monitor and the other on the preview monitor.

The video lines are looped through the distribution switcher, then fed to the production switcher. This is done because most active distribution switchers allow looping and most inexpensive production switchers have terminated inputs. The picture monitors and CROs are looped to eliminate the need for DAs. DAs are desirable, but would increase the cost of the system.

The audio system uses a small single-channel console. If one with the required number of inputs is not in your price range, a console with switch selected inputs could be used instead. For a system of this size, usually only a few high-level inputs are required for any one production. If the high-level inputs accept only -20 dbm levels, as is the case in some consoles, a 20 db pad can be used to cut down the level following the distribution switcher inputs.

Incidentally, if only -20 dbm inputs are available, a bridging transformer (20,000 ohms to 600 ohms) of good quality when bridged across a 0 dbm line will nicely match the -20-dbm input in both level and impedance when the 600-ohm side is connected to the console input. Inexpensive consoles providing high-level bridging inputs are hard to find, so such modifications can be useful. Of course, you could also pad down the input at the console and bridge at the distribution switcher, with termination at the console instead of the source. The Sparta Model A15-B, one possibility, has 600-ohm -10 dbm

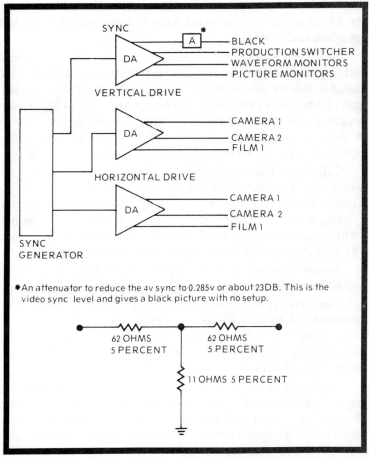

Fig. 15-7. System 1 pulse flow diagram.

high-level inputs, so the audio lines would best be bridged by the distribution switcher, possibly a Dynair VS-6X6C-SB, and terminated through a 10 db pad at the console.

Some distribution switchers might also require less than 0 dbm level. This would be noted on the specification sheet. Pads of suitable size would then be required. The distribution switcher used, obviously, should have audio follow and video and audio bridging inputs.

The audio console is located so it can be used by either system operator or a third person, if enough personnel are available.

The two VU meters in rack 1 monitor the test output of the distribution switcher and the audio console program output, so the switcher operator does not have to turn around to see the program audio level.

Small PA amplifiers are used to amplify the studio and test outputs of the distribution switcher. Any speakers of suitable quality can be used. Remember, the studio speaker must be muted when the microphones are open. The console program and cue speakers are driven directly by the console in some cases; others require external power amplifiers. Check the specifications.

Tally lights are operated by the spare contacts in the production switcher and are supplied by a 24-volt power supply. This supply can also be used to power the intercom system, if one is not provided. If the studio cameras do not include intercom, an external system will have to be provided. The distribution switcher is not tallied in this system.

Daven, among others, makes interphone amplifiers suitable for use in this system. They are very easy to hook up.

The sync system is a simple one, using a small EIA generator. A Dynair SY-290A generator with PD-241A pulse DAs could be used to supply the sync, vertical drive and horizontal drive. Blanking probably will not be used. If it is, an additional DA can be used to distribute it.

I will not presume to choose cameras and pedestals or tripods for this system; that selection must be determined by the quality you desire and the money you have to spend. Remember, you need driven cameras with viewfinders, tally lights and intercom capability for a system like this one. The lens selected will be determined by the size of your studio.

Typical RF distribution equipment for this system could be Dynair TX-2A modulators followed by the Dynair CF-200A channel amplifiers of the correct frequency. More expensive modulators are available, but the above modules were selected to keep the system price down. Again, the budget must be your guide.

A good tuner, widely used in television systems for air monitoring and recording off the air, is the Conrac AV12E. If you do record off the air, be sure to get permission from the originating station before using it.

The Telemation TMM-203 series is a possible choice for the film chain. The projector could be a Graflex 930 or Bell and Howell 614-EVMS or 614-ETVS. These can be purchased from Telemation with the proper lenses for use with the TMM-203 multiplexer. Also available from Telemation is the Kodak EK-650 slide projector modified for use with the TMM-203. Almost any driven vidicon remote control camera can be used with the film chain, such as the Telemation TMC-2100. No matter whose equipment you buy, be sure you get the proper lenses and mounting hardware for the camera and projectors. All equipment should be remote controllable. The film projector should work with the type of film sound you use, either magnetic or optical.

When you start checking prices of film chains you might decide to either forego film and slides or just use a uniplexer for film only. Such a unit is the Telemation TMU-103. Rather than using slides, if you have an art or graphics department, cards can be made for open and closing titles, credits and supers. Another possibility is the projection of slides or films on a front or rear screen in the studio for pickup by the studio cameras. This will work, but a uniplexer is a better method.

Suitable turntables are the QRK 12S-1 or 12S-4, Rek-O-Kut B-12H, or other similar units with the recommended tone arm. You have a wide choice of preamps, or they might be built into the audio console. In any case, the preamp output impedance and level should match that required for the audio console you choose.

The ATR can be a monaural high fidelity type machine of good quality. It will probably have to be customized a bit to mount in a rack. Transformers will probably be required for the input and output to match the system impedances.

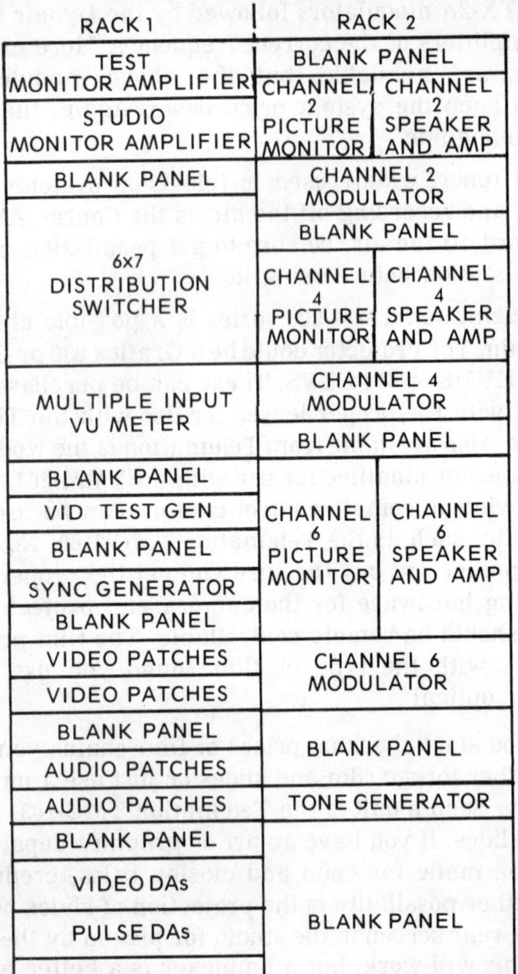

RACK 1	RACK 2
TEST MONITOR AMPLIFIER	BLANK PANEL
STUDIO MONITOR AMPLIFIER	CHANNEL 2 PICTURE MONITOR / CHANNEL 2 SPEAKER AND AMP
BLANK PANEL	CHANNEL 2 MODULATOR
6x7 DISTRIBUTION SWITCHER	BLANK PANEL
	CHANNEL 4 PICTURE MONITOR / CHANNEL 4 SPEAKER AND AMP
MULTIPLE INPUT VU METER	CHANNEL 4 MODULATOR
	BLANK PANEL
BLANK PANEL	CHANNEL 6 PICTURE MONITOR / CHANNEL 6 SPEAKER AND AMP
VID TEST GEN	
BLANK PANEL	
SYNC GENERATOR	
BLANK PANEL	
VIDEO PATCHES	CHANNEL 6 MODULATOR
VIDEO PATCHES	
BLANK PANEL	BLANK PANEL
AUDIO PATCHES	
AUDIO PATCHES	TONE GENERATOR
BLANK PANEL	BLANK PANEL
VIDEO DAs	
PULSE DAs	
24v POWER SUPPLY	

Fig. 15-8. Rack layout for System 2.

We have tried to make this as simple and inexpensive a system as possible, only using those components we felt necessary. The more complex systems described in this chapter may suggest additional features to improve this system.

The monitor and remote control placement is a little inconvenient, but is the best possible using that number of

RACK 3		RACK 4		RACK 5	
BLANK PANEL		ATR		BLANK PANEL	
CAMERA 1 PICTURE MONITOR	CAMERA 2 PICTURE MONITOR	FILM CRO	FILM PICTURE MONITOR	PROGRAM PICTURE MONITOR	TEST PICTURE MONITOR
TALLY LIGHT		TALLY LIGHT		BLANK PANEL	
BLANK PANEL		REMOTE CONTROLS		PROGRAM AND TEST CRO	
CAMERA 1 CRO	CAMERA 2 CRO			BLANK PANEL	
				6 x 1 TEST SWITCHER	
BLANK PANEL		BLANK PANEL		BLANK PANEL	
SHELF		SHELF		SHELF	
BLANK PANEL		BLANK PANEL		BLANK PANEL	

monitors. You may be able to change things around to suit your operation.

Parts of the previously described system can be used alone; for example, perhaps you might want to use only two camera systems, a pulse system, switcher, one VTR and modulator along with enough monitors to operate the system. In this case, a smaller audio console, such as the Shure M67, could be used. This would allow simple 2-camera productions. A distribution switcher would then not be needed and the VTR could feed the modulator directly, with black derived from the production switcher. The studio video monitor could be looped from the switcher program out and the studio audio could come from the audio console.

Prepackaged systems of this size are available too. Some are designed to be portable, others are intended for use in one area. Such prepackaged systems are designed for production, so you will have to add the distribution parts yourself, if they are required.

SYSTEM 2

This system is essentially the same as the small system described previously, but has some additions which should prove useful. Diagrams are shown in Figs. 15-8 through 15-13.

Let's begin with the production switcher. This could be a switcher similar to that used in the small system with a special effects feature added (don't forget the delays), or an integrated system such as the Dynair VS-153A, which allows wipes and keys or mats to be used. Remember, any special effects unit can be used with any switcher as long as the necessary switching and delays are provided. For example, Dynair and Ball Brothers, among others, manufacture special effects units that can be added to any switcher. If switching to select the input to the special effects is not built-in, distribution switchers can be added. Nonsynchronous inputs need not be run to the effects, since they cannot be used.

The switcher mentioned has the effects built-in and will accept all the inputs shown in the system block diagram. Note: if this unit is selected, all the synchronous inputs must be either composite or noncomposite; they cannot be mixed. This should present no problems, however.

Monitors for all the production switcher inputs can be put on shelves in front of the switcher position, along with program and preview monitors. No monitor is provided for the test generator because it shouldn't be needed. Nor is the test generator applied to the effects. If necessary, it could be patched into the unused effects input.

The distribution switcher shown is a locally controlled model such as the Dynair VS-6X7C-SB. A separate test switch-

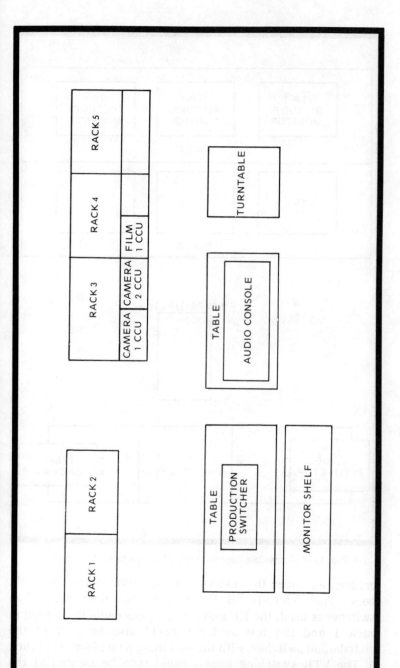

Fig. 15-9. System 2 control room floor layout.

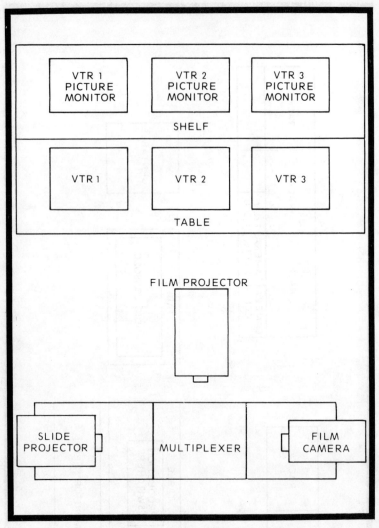

Fig. 15-10. A possible floor diagram of the System 2 telecine room.

er, located under the test switcher monitor in rack 5, could be a Dynair VS-6X1C-SB. If a remote controlled distribution switcher is used, the RF and studio banks could be located in rack 1 and the test switcher could also be part of the distribution switcher, with the switching panel located in rack 5. The VTR switching panels could then be located at the VTRs. This is more convenient but more expensive. Such a switcher could be a Telemation TAS-TVS - 12X series. Notice this model has 12 inputs rather than 6. As of this writing, to the best of my knowledge, no off-the-shelf remote control 6-input audio-video distribution switchers are available. Possibly something might be when you are assembling your system. In any case, the extra inputs would be valuable for future expansion. Another alternative would be a custom distribution switcher (not so expensive as you might think).

The film chain can be like the multiplexer-slide projector-film projector system outlined for the previous system.

All three VTRs are shown returning to the production switcher. Possibly you might want to purchase one top-of-the-

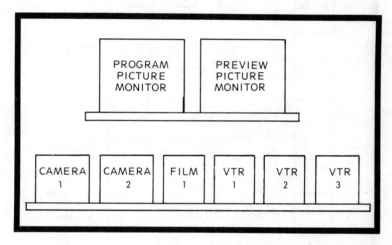

Fig. 15-11. The System 2 picture monitor layout at the production switcher.

line VTR of the format you have chosen, which can be locked to the house sync to provide another synchronous source. The less expensive VTRs, as you know, are nonsynchronous. For playback to the distribution switcher, synchronous sources are not necessary. If a synchronous VTR is used in the system, it should be routed through the effects.

The film chain and VTRs are shown in a separate room, though, of course, this is not necessary if the control room is large enough. Proc amps can be used with the VTRs if desired. If not built into the VTR they are a useful accessory because the video and sync levels can be adjusted and the instability noted in some TV receivers caused by the crossover dropout in single-head helical scan tape recordings is eliminated. The

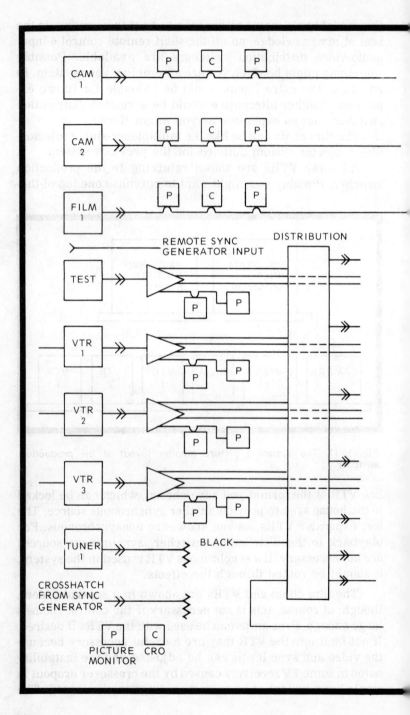

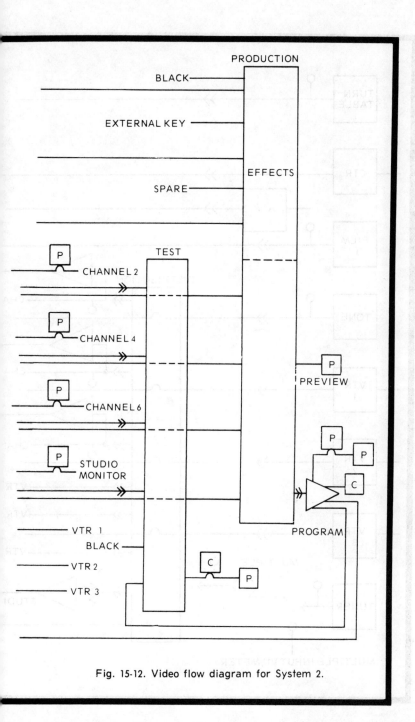

Fig. 15-12. Video flow diagram for System 2.

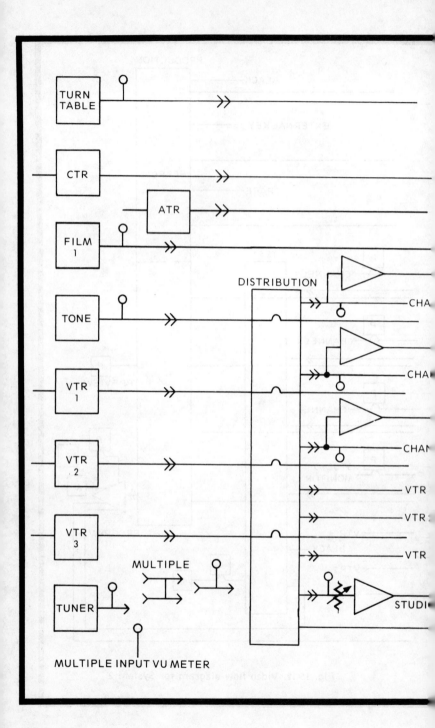

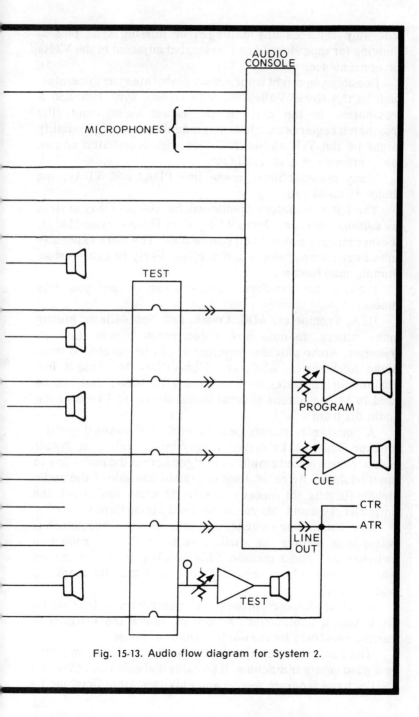

Fig. 15-13. Audio flow diagram for System 2.

proc amp replaces any damaged or missing sync pulses. Shelving for tape storage can be located adjacent to the VTRs for convenience.

Possibly you might want a more elaborate sync generator, such as the Grass Valley 950 with remote sync lock and a crosshatch. In the case of the Grass Valley unit, the crosshatch requires an additional module. Another possibility might be the Telemation TSG-2000 with crosshatch option. Many others are also available.

Many manufacturers make fine PDAs and VDAs; the choice is up to you.

The RF modulators mentioned for the previous system are suitable here, or a Jerrold CMC-A or Dynair Dyna-Mod II, or other more expensive unit can be used. The more expensive units have sharper skirts and are less likely to cause cross-channel interference.

Picture and waveform monitors can be any you may choose.

RCA, Trompeter, ADC, Cooke, and Nems-Clark, among many others, manufacture video patch panels and accessories. Audio patching equipment can be purchased from Audio Accessories, ADC, and Nems-Clark to name a few manufacturers. Gates Radio makes a VU meter that can be used to bridge across several lines, shown as VU-22 in the audio diagram.

A speaker is shown next to each RF channel picture monitor. This can be driven simply by a small 1- or 2-watt cheap transistorized amplifier bridged across the audio line to the modulator. Be careful not to ground one side of the audio circuit (if it is not already grounded) when you attach the amplifier. To avoid this you might need a transformer.

One distribution switcher output feeds the studio, which is better than feeding the studio monitor with the production switcher and audio console. This speaker should be muted when a studio mike is opened. Of course, a monitor amplifier will be needed.

Cartridge tape recorders are manufactured by Collins, RCA, Gates, Spotmaster and others. Check the features of various recorders for the one that fits your needs.

The audio tape recorder could be a professional machine or a good amateur machine. If possible it should run at 7½ and 15 IPS, have separate record and playback amplifiers and be capable of being remote controlled. Ampex, Scully and Magnecord, among others, make good machines. Sony, as well as a number of other manufacturers, make good amateur machines.

Fairchild makes a tone generator, the 6920SC-5, with five frequencies. It needs some method of mounting and a power supply.

Twenty-four volt power supplies are available from numerous manufacturers. Be sure the supply you choose is capable of the current required. A self-protected power supply will save replacing a lot of fuses.

The audio console could be the CCA "Executive," QRK-8, Collins 212V-1, or some other similar unit.

A pulse flow diagram for system 2 is not shown because the requirements vary with the equipment chosen. Simply run the required pulse outputs from the sync generator to pulse DAs through a patch panel. From there they are distributed to the equipment. Don't loop too many pieces on one DA output line, or the pulses will become degraded, and if the line has to be opened at any point, all equipment beyond that point will be deprived of the pulses.

Crosshatch from the sync generator should be brought out to a patch on the video patch panel; then it can be patched wherever desired.

The cameras indicated are monochrome, though, of course, color cameras could be used. If this is your desire, see the system described next. As in the small system, choose cameras with viewfinders, tally lights and intercom facilities. They must also be capable of being driven.

SYSTEM 3

This system is similar to the preceding systems, but with the addition of color and a few new items. Figs. 15-14 through 15-20 show all the details.

The production switcher might have two pairs of mix—effects banks, which would require an additional monitor in the production control area. The switcher used to feed the VTRs into the vectorscope and color monitor should be of excellent quality, so the phase of the signal will not be changed and distortion avoided.

It is assumed that VTR 1 and VTR 2 will be top-of-the-line machines, such as the IVC 900 series or the Ampex VPR7900.

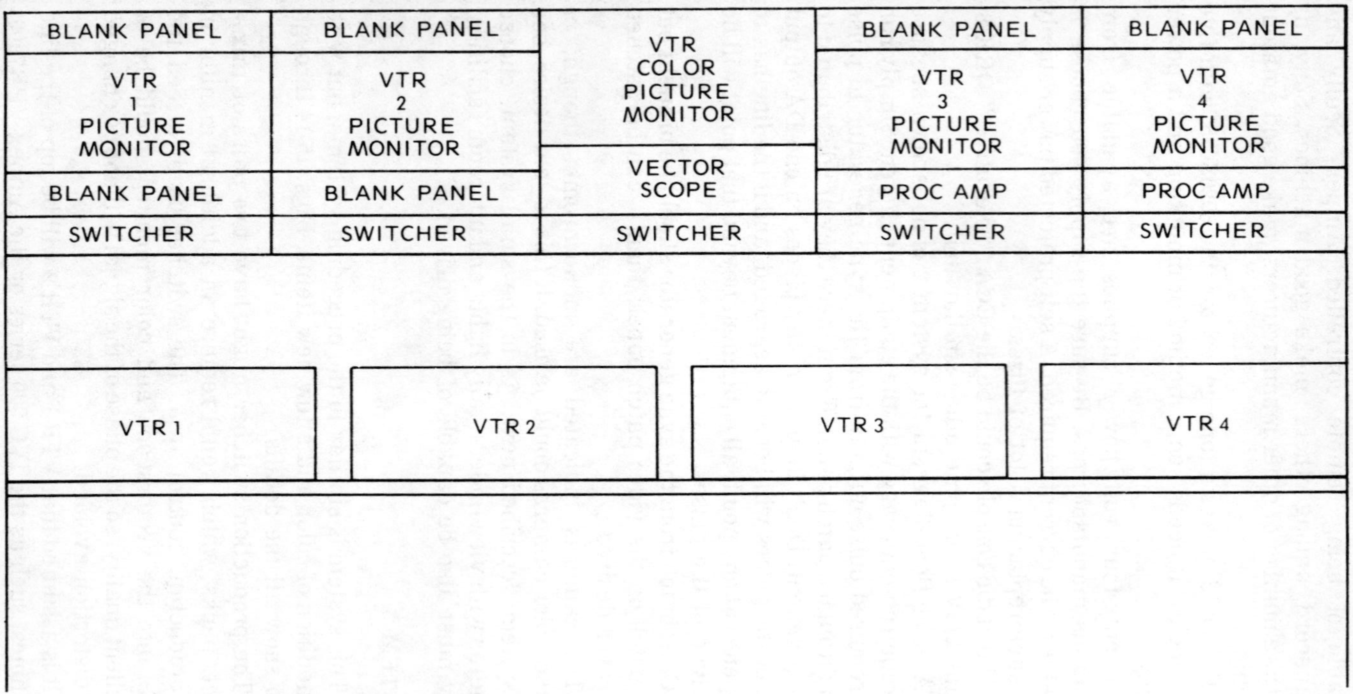

BLANK PANEL	BLANK PANEL		BLANK PANEL	BLANK PANEL
VTR 1 PICTURE MONITOR	VTR 2 PICTURE MONITOR	VTR COLOR PICTURE MONITOR	VTR 3 PICTURE MONITOR	VTR 4 PICTURE MONITOR
		VECTOR SCOPE	PROC AMP	PROC AMP
BLANK PANEL	BLANK PANEL			
SWITCHER	SWITCHER	SWITCHER	SWITCHER	SWITCHER

VTR 1	VTR 2	VTR 3	VTR 4

Fig. 15-14. System 3 VTR equipment layout.

They should include proc amps and drop-out compensators. The other two machines could be less expensive, but compatible with the recording format of VTR 1 and 2 of course. Proc amps are provided for VTRs 3 and 4; it is assumed they are part of VTR 1 and 2 and, therefore, not required as external accessories. All the VTRs could also be supplied with drop-out compensators if you wish.

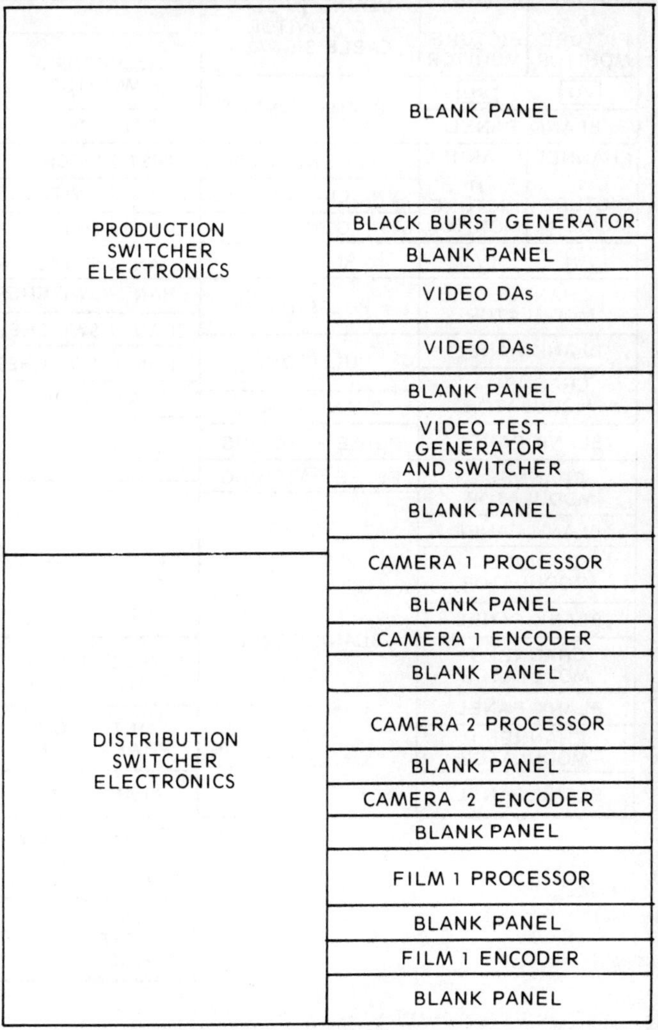

PRODUCTION SWITCHER ELECTRONICS	
	BLANK PANEL
	BLACK BURST GENERATOR
	BLANK PANEL
	VIDEO DAs
	VIDEO DAs
	BLANK PANEL
	VIDEO TEST GENERATOR AND SWITCHER
	BLANK PANEL
DISTRIBUTION SWITCHER ELECTRONICS	CAMERA 1 PROCESSOR
	BLANK PANEL
	CAMERA 1 ENCODER
	BLANK PANEL
	CAMERA 2 PROCESSOR
	BLANK PANEL
	CAMERA 2 ENCODER
	BLANK PANEL
	FILM 1 PROCESSOR
	BLANK PANEL
	FILM 1 ENCODER
	BLANK PANEL

Fig. 15-15A. Part of rack layout for System 3.

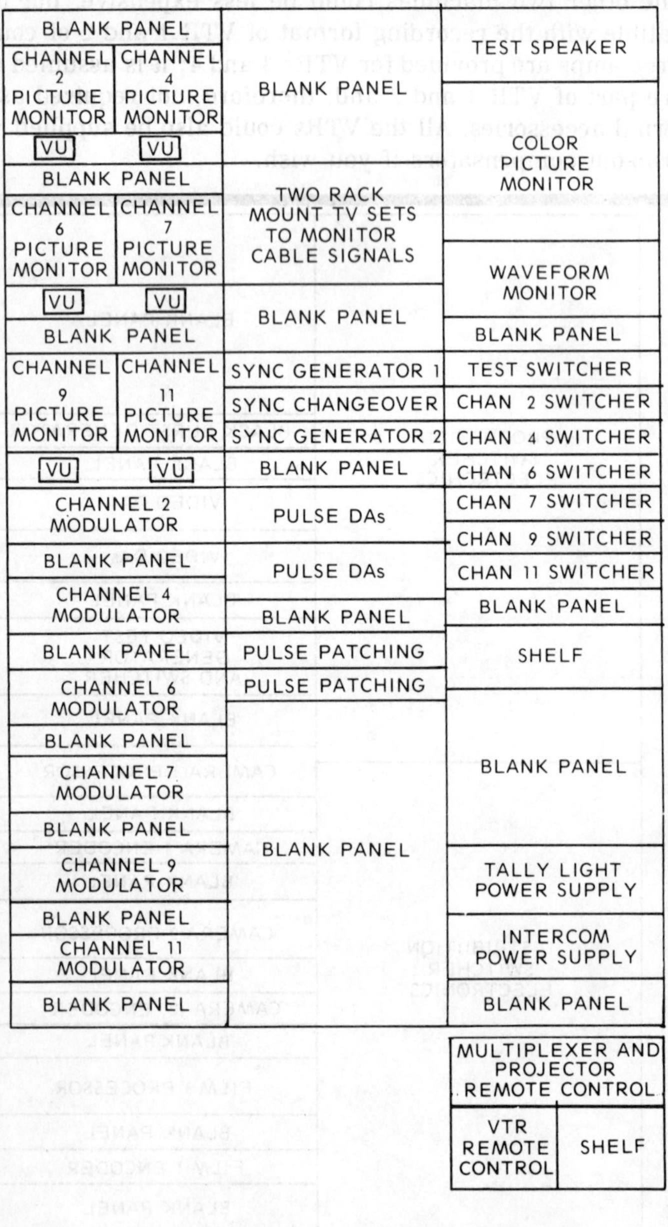

BLANK PANEL		BLANK PANEL	TEST SPEAKER
CHANNEL 2 PICTURE MONITOR	CHANNEL 4 PICTURE MONITOR		
VU	VU		COLOR PICTURE MONITOR
BLANK PANEL		TWO RACK MOUNT TV SETS TO MONITOR CABLE SIGNALS	
CHANNEL 6 PICTURE MONITOR	CHANNEL 7 PICTURE MONITOR		
VU	VU		WAVEFORM MONITOR
BLANK PANEL		BLANK PANEL	
BLANK PANEL			BLANK PANEL
CHANNEL 9 PICTURE MONITOR	CHANNEL 11 PICTURE MONITOR	SYNC GENERATOR 1	TEST SWITCHER
		SYNC CHANGEOVER	CHAN 2 SWITCHER
		SYNC GENERATOR 2	CHAN 4 SWITCHER
VU	VU	BLANK PANEL	CHAN 6 SWITCHER
CHANNEL 2 MODULATOR		PULSE DAs	CHAN 7 SWITCHER
			CHAN 9 SWITCHER
BLANK PANEL		PULSE DAs	CHAN 11 SWITCHER
CHANNEL 4 MODULATOR		BLANK PANEL	BLANK PANEL
BLANK PANEL		PULSE PATCHING	SHELF
CHANNEL 6 MODULATOR		PULSE PATCHING	
BLANK PANEL			BLANK PANEL
CHANNEL 7 MODULATOR			
BLANK PANEL		BLANK PANEL	
CHANNEL 9 MODULATOR			TALLY LIGHT POWER SUPPLY
BLANK PANEL			INTERCOM POWER SUPPLY
CHANNEL 11 MODULATOR			
BLANK PANEL			BLANK PANEL
			MULTIPLEXER AND PROJECTOR REMOTE CONTROL
			VTR REMOTE CONTROL / SHELF

Fig. 15-15B. Part of rack layout for System 3.

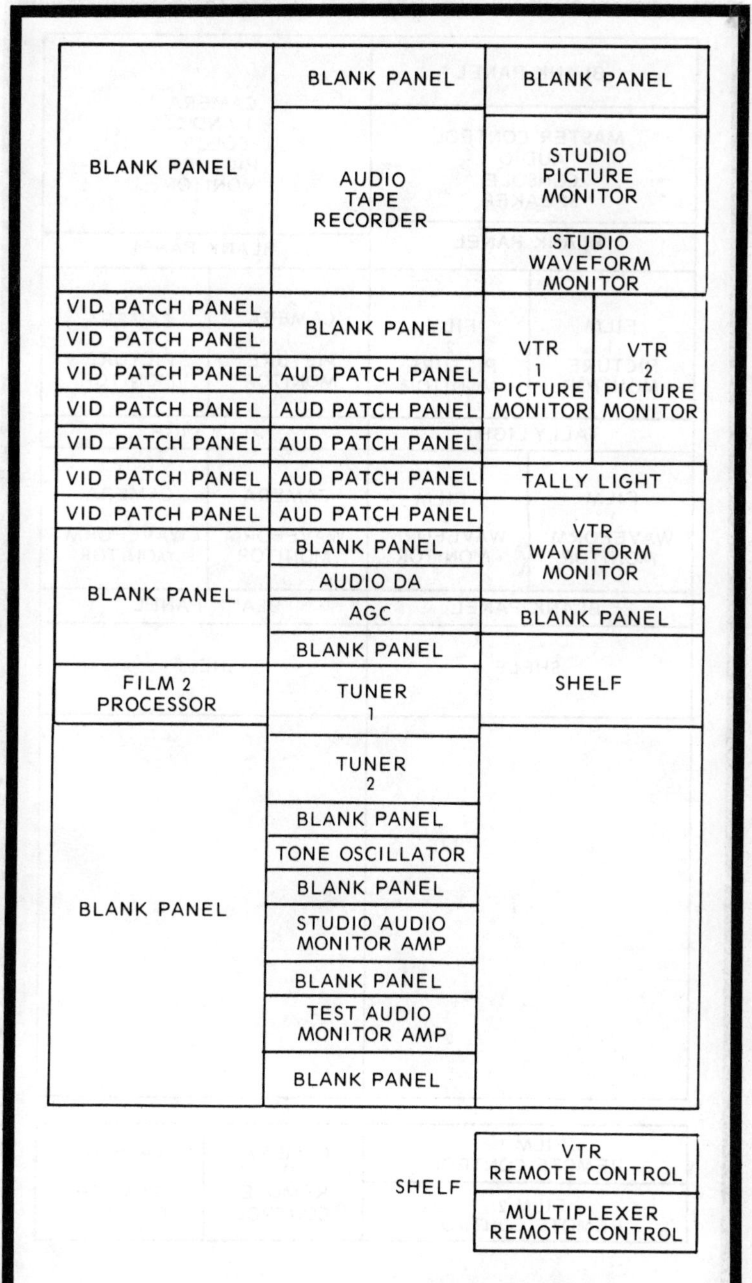

	BLANK PANEL	BLANK PANEL	
BLANK PANEL	AUDIO TAPE RECORDER	STUDIO PICTURE MONITOR	
		STUDIO WAVEFORM MONITOR	
VID PATCH PANEL	BLANK PANEL	VTR 1 PICTURE MONITOR	VTR 2 PICTURE MONITOR
VID PATCH PANEL			
VID PATCH PANEL	AUD PATCH PANEL		
VID PATCH PANEL	AUD PATCH PANEL		
VID PATCH PANEL	AUD PATCH PANEL		
VID PATCH PANEL	AUD PATCH PANEL	TALLY LIGHT	
VID PATCH PANEL	AUD PATCH PANEL		
BLANK PANEL	BLANK PANEL	VTR WAVEFORM MONITOR	
	AUDIO DA		
	AGC	BLANK PANEL	
	BLANK PANEL		
FILM 2 PROCESSOR	TUNER 1	SHELF	
BLANK PANEL	TUNER 2		
	BLANK PANEL		
	TONE OSCILLATOR		
	BLANK PANEL		
	STUDIO AUDIO MONITOR AMP		
	BLANK PANEL		
	TEST AUDIO MONITOR AMP		
	BLANK PANEL		

	SHELF	VTR REMOTE CONTROL
		MULTIPLEXER REMOTE CONTROL

Fig. 15-15C. Part of rack layout for System 3.

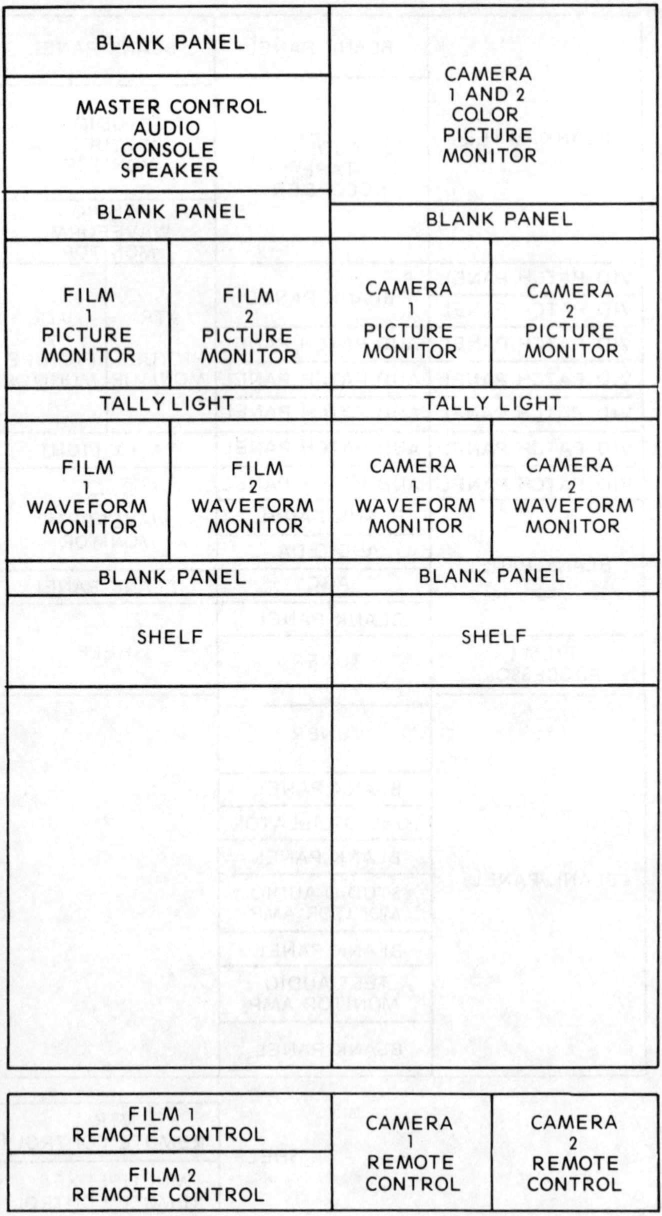

BLANK PANEL	CAMERA 1 AND 2 COLOR PICTURE MONITOR
MASTER CONTROL AUDIO CONSOLE SPEAKER	
BLANK PANEL	BLANK PANEL
FILM 1 PICTURE MONITOR / FILM 2 PICTURE MONITOR	CAMERA 1 PICTURE MONITOR / CAMERA 2 PICTURE MONITOR
TALLY LIGHT	TALLY LIGHT
FILM 1 WAVEFORM MONITOR / FILM 2 WAVEFORM MONITOR	CAMERA 1 WAVEFORM MONITOR / CAMERA 2 WAVEFORM MONITOR
BLANK PANEL	BLANK PANEL
SHELF	SHELF
FILM 1 REMOTE CONTROL / FILM 2 REMOTE CONTROL	CAMERA 1 REMOTE CONTROL / CAMERA 2 REMOTE CONTROL

Fig. 15-15D. Part of rack layout for System 3.

Only one film camera is shown to be a color camera. If a multiplexer similar to the Telemation TMM-211 or Sarkes Tarzian MP-90 is used, two film projectors and two slide projectors can be switched into either of the two cameras. Material to be recorded or distributed in color can be fed to the color camera; other material such as super slides, etc., can be fed to the monochrome camera. Of course, you can use two color cameras if you wish.

Two color studio cameras are shown, and a third can be added if necessary. If you add the third camera, you will have to add the DA and additional monitors required.

A useful accessory not shown on the system diagram would be an editing switcher. If you plan to do a lot of editing you could get a simple little switcher, with effects if desired, and bring three or four switcher inputs to the video patch bay. The edit switcher output can be fed into the distribution switcher. You could probably get by with only two monitors—preview and program. Be sure the switcher has good color specifications. A small audio mixing console will be required, such as the Shure M67 mixer or Ampex AM-10 mixer with accessory meter panel if you want a rack-mounted mixer. If a table top unit is acceptable, you could use the Collins 212J-1, Gates Producer or a similar types. The mixer inputs are brought out to the audio patch bay, with the output also fed to the distribution switcher. Using this additional equipment you can edit while the big switcher and audio console are being used for something else. It is also convenient to have the editing switcher and audio mixer close by the VTR.

A drawing of the pulse system has been omitted because the pulse requirements vary with the equipment selected. When using color, remember that the pulse timing is very critical. The sync generator will have to supply two additional signals—burst flag and subcarrier. DAs will have to be provided for these signals, too. The sync generator should have a remote input and crosshatch output. In a system this size, you might consider a back-up generator and a manual or automatic sync changeover switch. You could use either DAs or an encoder-decoder distribution system. Decoders with built-in adjustable delays make the timing adjustments easier. Sarkes Tarzian makes suitable units.

A great many companies manufacture production switchers suitable for such a system; Grass Valley, Cohu, Sarkes Tarzian, RCA and many others. A switcher this size is

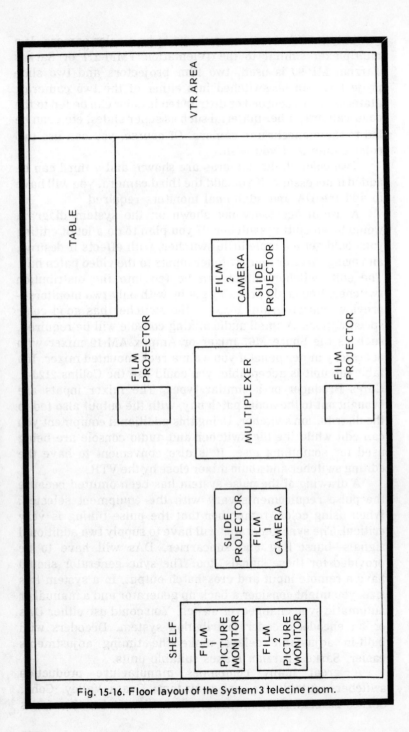

Fig. 15-16. Floor layout of the System 3 telecine room.

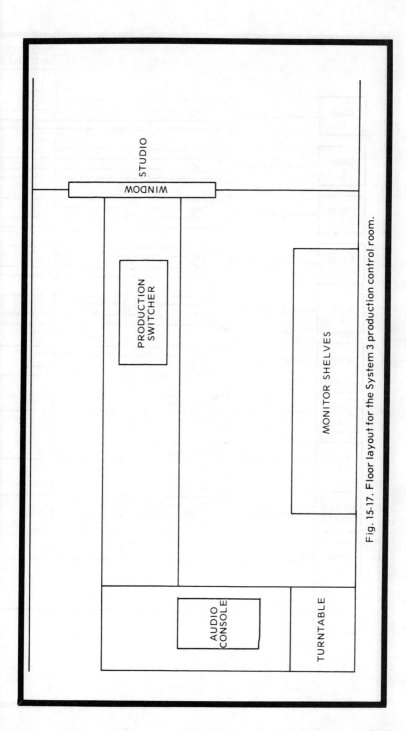

Fig. 15-17. Floor layout for the System 3 production control room.

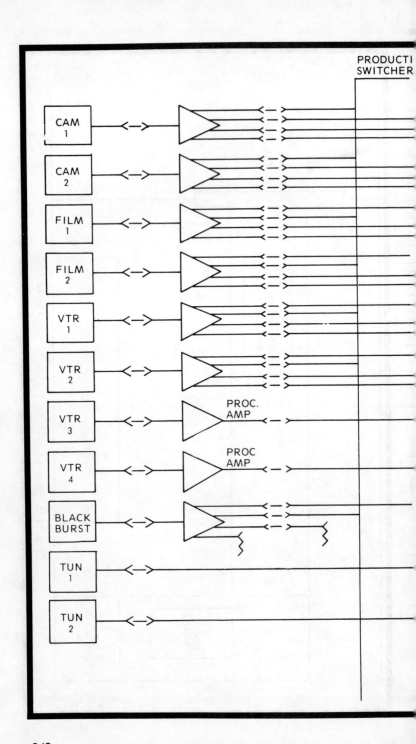

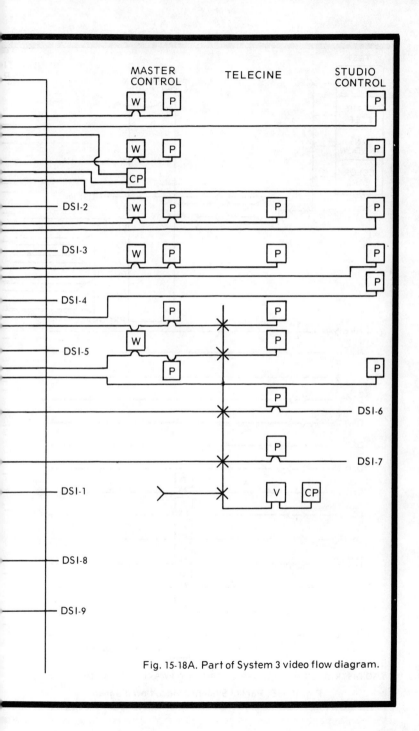

Fig. 15-18A. Part of System 3 video flow diagram.

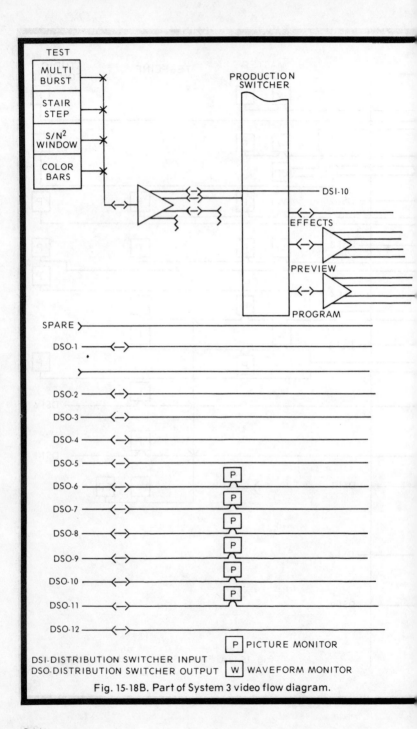

Fig. 15-18B. Part of System 3 video flow diagram.

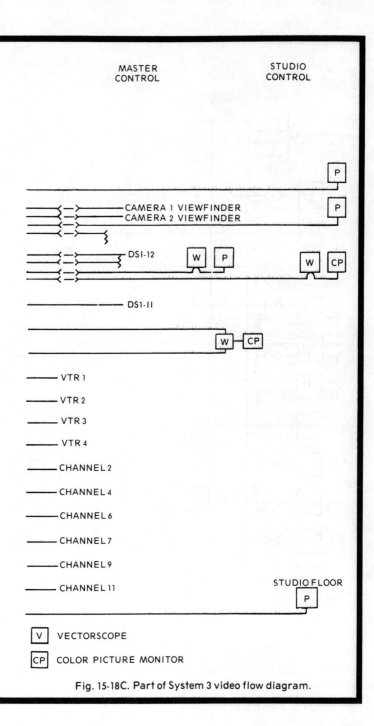

Fig. 15-18C. Part of System 3 video flow diagram.

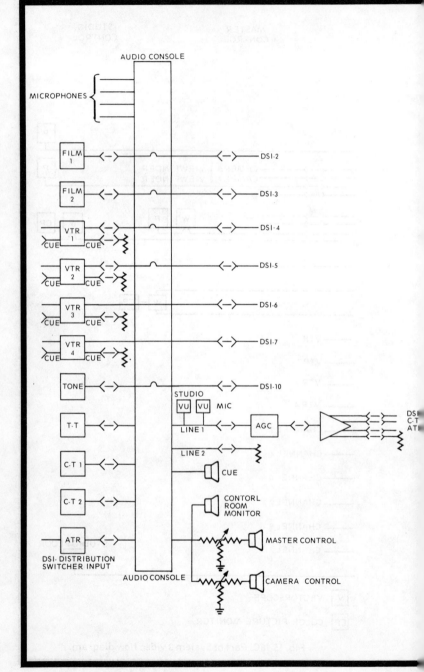

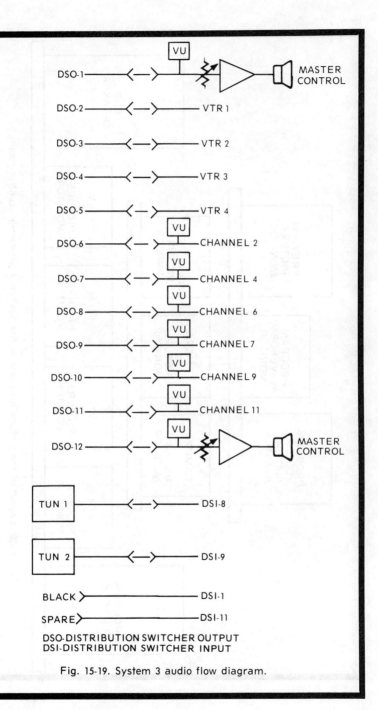

Fig. 15-19. System 3 audio flow diagram.

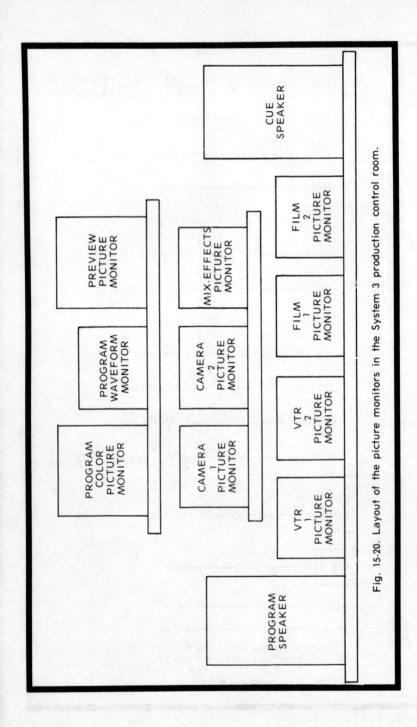

Fig. 15-20. Layout of the picture monitors in the System 3 production control room.

usually custom made, so be sure to ask for exactly what you want.

Two tuners are included. One can be connected to the distribution system to allow monitoring the signals you are distributing. The other tuner can be connected to an antenna or your community's cablevision feed.

An AGC amplifier is shown on the audio console output. Here again you have a wide variety of choices—Allison Research or CBS Laboratories units are possibilities.

Adjustable T pads are shown at the audio console remote speakers in master control. This is to allow them to be turned down when you are not listening to them.

The table shown next to the VTR area is handy to hold empty tape boxes, store tapes in use that day, etc.

The cartridge tape units are located under the turntable. Many suppliers, such as Gates Radio, offer pedestals designed to accommodate a turntable on top with 19-inch rack space in the side beneath the turntable.

A vectorscope is shown in the VTR area, since it will probably be used there more often than anywhere else. An additional patch input to the vectorscope switcher is provided so that any other signal can be examined with it.

The audio console could be a Collins 212T-1 or 212T-2, Gates "Diplomat," Norelco MD series, or one of many off-the-shelf or custom consoles.

In this and the previous two systems, any equipment shown on the diagrams but not described is covered in earlier chapters.

Index

A

Amplifier, processing, 127
Amplifiers, 172
Amplifiers, monitor, 148
Amplifiers, RF, 139
Animation, 60
Audio flow diagram, 215
Audio flow diagram for system 1, 216
Audio mixing consoles, 143
Audio processors, 171
Audio program sources, 155

B

Booms, microphone, 165
Bridging inputs, 28

C

Cable, long runs, 133
Cable, RF, 139
Cable, short runs, 133
Cable types, 133
Camera linearity adjustments, 110
Camera mounting, 43
Camera, 35
Camera setup, 42
Cartridge tape machines, 157
Cassette machines, 158
Chain, the film, 57
Color equipment, 93
Color monitors, 98
Color, VTR, 97
Component tests, 188
Composite sync, 10
Consoles, audio mixing, 143
Consoles, typical, 143
Controls, remote, 59, 180
Couplers, directional, 137

D

Design, systems, 202, 209
Diagram, floor, 213
Directional couplers, 137
Disc playback equipment, 159
Distribution, pulse, 15
Distribution switcher, 28
Distribution system, 133
Distribution system, RF, 135
Dolby circuitry, 156
Drawing symbols, 203

E

Equipment, color, 93
Equipment, peripheral audio, 169
Equipment racks, 178
Equipment racks system drawing, 212
Equipment, special, 61
Equipment, telecine, 57
Equipment, test, 140

F

Film chain, 57
Floor diagram, 213

G

Generators, pulse, 9
Generators, signal, 187

H

Helical VTRs, 84
Horizontal sync, drive pulse diagram, 14

I

Inputs, bridging, 28
Inputs, terminated, 28
Intercom systems, 176
Interlace sync, random, 14

L

Lighting, 49
Lighting equipment, 35
Linearity adjustments, camera, 110

M

Machines, cassette, 158
Maintenance shop, 184
Maintenance tips, 194
Methods, switching, 21
Microphone booms, 165
Microphones, 161
Microphone stands, 165
Mixing consoles, audio, 143
Monitor amplifiers, 148
Monitors, color, 98
Monitors, RF, 105
Monitors, video, 105
Monitor, TV projector, 113
Monitor, waveform, 115
Motion picture projectors, 62
Mounting, camera, 43
Multiplexers, 67

O

Operation, systems setup, 69
Oscilloscopes, 186

P

Pads and splitters, 176
Panels, patch, 129, 169
Parts, spare, 191
Passive devices, other, 137
Passive devices, splitters, 137
Patch panels, 129, 169
Patterns, test, 46
Peripheral audio equipment, 169
Playback equipment disc, 159
Processing amplifier, 127
Processors, audio, 171
Production switcher, 26
Program sources, audio, 155
Program switcher, 28
Projectors, motion picture, 62
Projectors, slide, 58
Pulse distribution, 15
Pulse generators, 9
Pulse train diagram, 13
Putting a system together, 198

Q

Quadrature VTRs, 76

R

Rack layout for system 2, 222
Racks, equipment, 178
Random interlace sync, 14
Recorders, video, 73
Recorder, video cartridge, 88
Recorder, video disc, 90
Remote controls, 59, 180
Retma resolution chart, 46
RF amplifiers, 139
RF cable, 139
RF distribution system, 135
RF monitors, 105

Servo, circuitry, 77
Setup, camera, 42
Shop, maintenance, 184
Signal generators, 187
Slide projectors, 58
Spare parts, 191
Special effects, 21
Special effects unit, 31
Special equipment, 61
Specifications, 198
Splitters and pads, 176
Splitters, passive devices, 137
Stands, microphone, 165
Studio equipment, 54
Studio pulse system, 7
Switcher, distribution, 28
Switcher, production, 26
Switcher, program, 28
Switchers, typical, 23
Switching methods, 21
Switching systems, 21
Sync, composite, 10
System designs, 202, 209
System, distribution, 133
System 1 pulse flow diagram, 219
System 2, 224
System 2 control room floor layout, 225
System 2 picture monitor diagram, 227
System setup and operation, 69
Systems, intercom, 176
System 2 telecine floor diagram, 226
System 2 video floor diagram, 229, 230
System 3, 233
System 3 production control layout, 241
System 3 production control room, 248
System 3, rack layout, 235-238
System 3 telecine floor layout, 240
System 3 video flow diagram, 243-247
System 3 VTR equipment layout, 234

T

Tape machines, cartridge, 157
Telecine equipment, 57
Terminal equipment, video, 121
Terminated inputs, 28
Test equipment, 140
Test generator, video, 122
Test patterns, 46
Tests, component, 188
Test signal, vertical interval, 126
Tips, maintenance, 194
Tools, 193
Train diagram, pulse, 13
Turntables, 160
TV projector monitor, 113
Typical consoles, 143
Typical switchers, 23

V

Vertical interval test signal, 126
Video cartridge recorder, 88
Video disc recorder, 90
Video monitors, 105
Video recorders, 73
Video terminal equipment, 121
Video test generator, 122
VTR, color, 97
VTRs, helical, 84
VTRs, quadrature, 76

W

Waveform monitor, 115